T0316844

Rok Ovsenik
Iwona Kiereta
(eds.)

Destination Management

PETER LANG

Frankfurt am Main · Berlin · Bern · Bruxelles · New York · Oxford · Wien

Bibliographic Information published by the Deutsche Nationalbibliothek
The Deutsche Nationalbibliothek lists this publication in the Deutsche Nationalbibliografie; detailed bibliographic data is available in the internet at <http://www.d-nb.de>.

Cover design: Atelier Platen

ISBN 3-631-55314-5
US-ISBN 0-8204-9920-X

© Peter Lang GmbH
Europäischer Verlag der Wissenschaften
Frankfurt am Main 2006
All rights reserved.

All parts of this publication are protected by copyright. Any utilisation outside the strict limits of the copyright law, without the permission of the publisher, is forbidden and liable to prosecution. This applies in particular to reproductions, translations, microfilming, and storage and processing in electronic retrieval systems.

Printed in Germany 1 2 3 4 6 7

www.peterlang.de

Destination Management

Preface

Times are changing; fiercer competition, uncertainty and complexity coupled with ever less ability to anticipate developments have created dramatic changes in the environment in terms of deregulation, mass takeovers and the necessary integration of supply chains... These changing times are especially impacting both the service sector in general – and tourism in particular.

In order to respond to the increased demand in terms of developing destination management by adopting new approaches, new solution models including new systems and in this way adapting to the changes in the environment, organizations are being obliged to conduct more in-depth research and implement new strategies.

The monograph presents fifteen contributions, which represent new insights and reveal possible options for adopting a new approach in the area of tourist destinations. After examining the scientific contributions readers will discover that destination management in Slovenia is the subject of frequent complex research. Researchers are looking for responses to open-ended questions, which are already clearly shown in this book.

The book provides a complete thematic examination of destination management – from paradigm questions, thoughts about relations, the new "third route" in tourism, insights into the methods in evaluating intellectual capital to above all marketing models for solutions and examples of good practice. Although this monograph represents a thorough examination of destination management in the region of Slovenia, it will also be an interesting read for destination management researchers from different parts of the world.

Slovenia is a small European country of approximately 20.256 km^2, inhabited by 2.2 million people. It is located in the heart of Europe, only 130 km from Venice, 200 km from Salzburg, 250 km from Vienna and 280 km from Budapest. It is situated at the crossroads of the Alpine region, Panonian plain, Dinaric mountains and the Adriatic sea.

Slovenia was a part of Yugoslavia for almost 60 years and gained independence in 1991 in the wake of the disintegration of Yugoslavia. Slovenia was the most developed republic in former Yugoslavia and is now the most developed among the newly formed countries in the territory of former Yugoslavia. In the year

2000 it had a GNP of € 16,000 per capita, which was almost 70% of the European Union average, while from 1995 to 2001 the average growth of GNP was approximately 4.5% per year, well above the European average (Pesci, 2003). In comparison, Croatia, which ranks second on the development scale among the former Yugoslav republics, had a GNP of € 6,000 per capita in the year 2000.

Tourism is one of the most important industrial branches in Slovenia; it directly or indirectly employs 52,000 people, almost 6.5% of the total labour force. Tourism accounts for 3.5% of GNP, whereas tourism, the catering industry and the travel industry together account for approximately 9.1% of GNP. In terms of exports, goods and services exported by the tourist industry in Slovenia account for 10%.

In the year 2000 Slovenia was visited by at least 2,158,240 tourists. Every year about 1.8 million people visit the top twenty Slovenian historic and natural sights and cultural monuments. The number one natural sight is the Postojna Cave in the Karst region. Other places are also well visited: the Savica Falls, Škocijan Cave, the Maribor Aquarium, the Volčji Potok arboretum, as well as the museum in the town of Bled, the Lipica stud-farm, Ptuj Museum, Bistra Engineering Museum, the Kostanjevica at Krka Art Gallery, the Church and ancient pharmacy in Olimje, and the Old Town in Celje, to name a few.

Turistica – College of tourism Portorož is a member of the youngest University in Slovenia, the University of Primorska. The college itself was formed in 1995. The founders were Slovenian tourism companies, and the college's mission was to establish a system for educating and training personnel for Slovenia's tourism industry. The college curriculum is internationally comparable, since it was designed in cooperation with the European PHARE program (TEMPUS).

Turistica is the only college in Slovenia to provide education in the field of tourism at university level. Its basic mission is to provide university-level education and programs for the acquisition of various professional licenses in the tourism industry and to organize several academic and industry conferences and other forms of knowledge exchange.

One of Turistica's missions is research. In pursuit of this goal, Turistica is engaged in numerous developmental and research projects in Portorož, the biggest tourism centre in Slovenia.

This book reflects the diversity of research interests of the academics involved in education and research at Turistica, College for Tourism in Portorož, Slovenia.

Rok Ovsenik, Boris Bukovec and Janja Jerman are the authors of the paper "Managers in the Slovenian Tourist Trade Industry and Their Attitude towards Their Country's Admission into the European Union". They are interested in the attitudes of Slovenian tourism managers towards one of the most important political events after Slovenia's separation from Yugoslavia. The authors stress that this was a significant social step for Slovenia in many areas where managers were prepared for the change. But there still were some significant differences. According to authors, tour operators have been preparing more intensively than managers from other parts of the tourism industry. In general, the authors' empirical research reveals that most managers do not fear or have any reservations about Slovenia's admission to the European Union.

Milan Ambrož is contributing to the book with his analysis of the new approach to the tourism industry, labeled by Burns as the "Third" one. Ambrož' paper reveals that evenly distributed social power is critical for the development of the tourism industry. The most interesting finding of the empirical research conducted by Milan Ambrož was that "a strong tendency for local communities (in Slovenia) to control development could hinder the entrepreneurial initiative of the private tourism sector".

Gorazd Sedmak's paper "Differentiation of Catering Outlets as a Variable in Tourism Destination Positioning" analyses the role of local food as a destination attraction element. The author presents the findings from two different research projects conducted in Slovenia in 2000 and 2003. He had analyzed two tourist destinations that were typical mass tourism resorts in the past and found that restaurant managers still preferred to offer cuisine that was tailored for mass tourists, despite the fact that both destinations have changed significantly over the last decade. In contrast, the author's research clearly shows that tourists yearn for authentic experiences, and do not take the classic approach that was expected in the mass tourism era.

Maja Uran and Janja Jerman were interested in "Service Quality as a Competitive Advantage in the Slovenian Hotel Industry". They present research based upon a survey of 5000 questionnaires. From the theoretical standpoint they follow Parasuraman's service quality model. The INSQPLUS model used in this Slovenian research appears to be a good tool for measuring internal factors - internal service quality.

Tadeja Jere Lazanski presents a theoretical paper on "System Dynamics Models as Support for Decision-Making in Tourism". After presenting the basic concept of systems, system dynamics and their possible application to the tourism industry, the author demonstrates how System Dynamics Models could be used to aid tourism-related decision-making. One possible use is within the National Tourism Strategy of Slovenia.

In their paper "Fundamentals of the New Paradigm of Quality Organizational Change Management", Boris Bukovec and Rok Ovsenik present a variety of contemporary approaches and models for organizational change management. These approaches were researched using various methodologies presented in the paper. After presenting most impressive results, the authors conclude that striving for change is in essence the qualitative aspect of change and assert that the paradigm of change could be relabeled a "paradigm of quality change management".

In his paper "How to Evaluate Intellectual Capital", Franko Milost presents several non-monetary and monetary models for intellectual capital evaluation. The author's interest in intellectual capital evaluation stems from his opinion that it is essentially an economic asset. If so, any human resource management should utilize models for evaluating intellectual capital.

Gordana Ivanković claims in her paper that it is absolutely necessary to design a marketing-oriented accounting system in the hotel industry in order to be able to identify the level of achievement produced by the adopted strategy. Her statements are supported by research findings focusing on measuring the performance of Slovenian hotels. Her paper is titled "Marketing-Oriented and Strategic Management Accounting".

Živa Čeh is a linguist. The main issue in her paper "Non-native Speakers Communicating in English: The Language of Tourism" is the fact that the English language is used even when there are no native speakers of English involved in the conversation. She explains the use of numerous possible word combinations, contributing to a better understanding of the speaker's language.

Goran Vukovič, Jure Meglič and Branislav Šmitek offer a "Model of an Interactive Presentation for a Tourist Destination Offer". The model is designed on the basis of research conducted by the main Slovenian tourist organization, Tourist association of Slovenia and aligned to the strategy development guidelines for Slovenian tourism. In this respect the authors are promoting the use of contemporary information communication technology, which connects all the players in tourism and can be used as a powerful tool.

Marko Tkalčič and Matevž Pogačnik try to characterize a tourist adapted destination search system in their paper "Tourist Adapted Destination Selection". In their search for an appropriate model for destination selection they have found two essential conditions. The first is the number of tourists that are inclined to use modern technology to select their destination, and the second is that user's satisfaction with the service.

Emil Juvan, Rok Ovsenik and Goran Vukovič present contemporary tourist destination management from the standpoints of local residents who are looking for a stable and sustainable tourism economy that is ecologically and socially conscious. An analysis of the empirical data they collected in the Mislinja Valley leads them to conclude that only integrated operations and actions at both horizontal and vertical levels of the travel industry will enable small urban areas and small tourist destinations to follow contemporary trends of tourism demand in the future. Their paper is "Feasibility of Tourism Destination Management and its Development in Small Urban Areas – Case of the Mislinja Valley".

Bojan Vavtar is a lawyer who analyses several dimensions of the concept of management and definitions of the rights or legal obligations of management in his paper "Supervision as a Function of Management in a Joint Stock Company". He claims that the notion of management is not a legal concept. As a result of his analysis, he states that "it is possible to enforce a manager's liabilities as a responsibility in relation with his obligations, which are directly defined, as well as in relation with his responsibility for conditions and business decisions".

Marko Ferjan claims that many companies in Slovenia expend large amounts of money on educating their employees and expect proportional results from that education in his paper "A Model for the Education of Tourism Workers". According to his research, he found that adult participants are particularly influenced by the following elements of education: course content, educational materials, instructor and the approach to execution or implementation.

Margareta Benčič, Rok Ovsenik and Iztok Purič contribute the paper "Controversies and Perspectives of Destination Management". One apparent finding from their research is that the participants rated the development strategy of Slovenian destination management unfavorably, especially in terms of the clarity and definition of the goals. This has made the researchers aware of the importance of the Slovenian state improving its understanding and support of networking economic, profit and non-profit interests at destination level.

Rok Ovsenik, Ph.D., Iwona Kiereta, Ph.D.

Rok Ovsenik, Boris Bukovec, Janja Jerman

Managers in the Slovenian Tourist Trade Industry and Their Attitude towards Their Country's Admission into the European Union

Abstract

The article briefly portrays Slovenia, a small European country that has been admitted into the EU as a full member. It shows some of the most important economic and social aspects of the transition of a former socialist country, which take that process as a condition of adjustment to the community of the Western European countries. Attitudes of the managers employed in the Slovenian tourist industry towards their country's admission into the European Union have undergone empirical analysis. Most of the managers do not show any fear or reservations. Those managers employed in the tourist industry that have been exposed to foreign markets so far are more specific in that there are no statistical differences regarding the various gender and age groups of managers.

Key words: Transition, tourist industry, management, tourism management, Republic of Slovenia, European Union, restructuring processes

1. Introduction

1.1 A Few Words about Slovenia

Slovenia is a small European country situated at the crossroads of the following geographical regions: the Alps, Pannonian plain, Dinaric mountains and the

Adriatic Sea. It has an area of only 20,256 km² (approximately half the territory of Switzerland). There are the Alps with the Triglav National Park, one of the biggest in Europe (Pesci, 2003), numerous lakes and rivers, a short but well indented coastline and large woods, all in this small area. It has a total of only 2 million inhabitants and an average population density of approximately 100 inhabitants per km² (Koscak,1998: 81). The country has succeeded in maintaining a relatively high degree of development after the separation from the former constituent state Yugoslavia, so that in the year 2000 it had a national revenue of € 16,000 per capita, which was almost 70% of the European Union average and in the interval from 1995 to 2001 the average increase was approximately 4.5% per year, which was above the European average (Pesci, 2003).

The country is located in the heart of Europe, at a distance of 130 km from Venice, 200 km from Salzburg, 250 km from Vienna and 280 km from Budapest (Koscak,1998: 84). Ljubljana, the capital city, is only a little more than 100 km from the outskirts of the country in all directions.

The tourist trade industry is one of the most important industrial branches in Slovenia and has been rapidly growing in the past few years. According to the Statistical Office records, the tourist trade industry directly or indirectly employs 52,000 people, which is almost 6.5% of the total labor force. The tourist sector accounts for 3.5% of GDP, whereas tourism, the catering industry and the travel industry together account for approximately 9.1% of GDP. In terms of exports, goods and services exported by the tourist industry in Slovenia account for about 10%. Up until 2020 the tourist industry is forecast to grow by 6% a year (Central Statistical Office of the Republic of Slovenia, 2002). In 2002 all tourism related activities produced total revenue of €1.2 billion, which is about 10% of the total revenue of all service trades. In that year Slovenia was visited by 2,158,240 tourists, which is more than the total population of the country (Pesci, 2003).

The Slovenian tourist industry offered accommodation for approximately 25,000 visitors in hotels, 55,000 in other forms of accommodation (such as boarding houses, private rooms and apartments, tourist vacation resorts, lodgings, rural tourist farms), as well as a further 14,000 on campsites in 2002 (Central Statistical Office of the Republic of Slovenia, 2002). This accommodation space will surely act as a limiting factor in the growth dynamics of the tourist industry. However, a rather low occupancy rate is one of the important resources that is being counted on in that development [so that according to the records of the Central Statistical Office of the Republic of Slovenia, the occupancy rate in 2001 was not more than 48% and other forms of accommodation about 35% (Central Statistical Office of the Republic of Slovenia, 2002).

Every year about 1.8 million visitors visit the top twenty Slovenian beauty spots and cultural monuments. Number one is the karsts in the Postojna Cave, which has been a natural tourist phenomenon for over 180 years. More than 28 million visitors, mostly foreigners, have visited it in that period. Some other natural phenomena also have a high attendance rate: the Savica Falls, Škocijan Cave, the Maribor Aquarium, the Volčji Potok arboretum, as well as the Museum in the town of Bled, the Lipica stud-farm, the Ptuj Museum, the Bistra Engineering Museum, the Kostanjevica at Krka Art Gallery, the Church and pharmacy in Olimje, the Old Town in Celje and others. To sum up, Slovenia offers not only the seaside and mountain attractions in winter, but also numerous interesting natural and cultural monuments that have been attracting tourists' attention for years (Meglič, et al., 2005:142). According to some claims, Slovenia is well known to European Union inhabitants as a tourist destination. (Clarke et al., 2001:194).

Together with some other small countries in the world, Slovenia realized that it had to exploit certain technological advantages that were available, at least in some areas and, for example, Slovenian designers and IT specialists have achieved remarkable results in their representation on the World Wide Web (Brunn, Cottle, 1997:245).

However, when it comes to scientific literature about Slovenian tourism, there is not much, at least not in the reviews registered in the scientific database. Searching through the scientific database (Web of Science) we found two of those reviews: M.Koscak's article (1998:85) about rural tourism and farm development published in *Tourism Management* and Gosar and Roblek's article (Gosar&Roblek, 2001:134) in *Mitteilungen der Oestereichischen Geographischen Gesellschaft* about specific aspects of tourist development, which addressed certain problems in the development of rural regions.

2   The Tourist Industry and Transition Processes as a Preparation for Admission into the European Union

The process of admission into the European Union has always had very interesting associations for Slovenia both from an economic and political perspective. It developed simultaneously with the transition from socialism to capitalism. It is important to stress that the transition was probably less painful for Slovenia than for some other former socialist countries, because the socialist social order was far more liberal in Slovenia than in certain Eastern bloc countries.

Slovenia had much livelier and far deeper contacts with capitalist countries at economic, political, educational and personal levels. Residents of Slovenia have been freely traveling abroad for decades, certain forms of economic exchange with foreign countries have been remarkably well developed, so that numerous forms of economic life are not totally incompatible, especially the tourist trade. As we have already pointed out, the Postojna Cave has been attracting tourists continuously for 180 years, and most of its visitors are from foreign countries. The situation was mostly the same during socialism in Slovenia, when it was a part of Yugoslavia. That confirms the claim that the transition process is not about creating new maps in an "empty" social and economic area, but about modification, reshaping the institutions and the practice of central planning (Williams&Baláz, 2002: 39) individually in every country, in accordance with their reality, tradition and achievements in social development.

Certainly, the integration of the economies of Central and Southeast European countries into a global economy, particularly the European Union, is a central point of the domestic and foreign policy of those countries (Hall, 1998:429) and the tourist industry has an important function in this process. Some of those countries (Czech Republic, Hungary, Poland and Slovenia) have registered large growth in the number of tourist visitors, particularly in the last decade (Hall, 1998:425). The development in the tourist industry in Poland, one of the most active candidates for admission into the EU, is particularly interesting. In the period from 1990 to 1996 it accomplished growth of 471% and improved from 28th to 9th place in the world's top ranked tourist destinations (Langlois et al., 1999:465). The Czech Republic, as a traditionally tourist country, has also achieved exceptionally good results in the post socialist period (Hall, 1998:429; Baláz, 1998:435.)

According to some authors, in the process of transition and admission into the European Union, the tourist trade is the one economic sector that has the necessary characteristics to make it a perfect example for: a) the development, privatization and flexibilization of services areas, b) the reduction of the centralization and control of bureaucracy, c) focusing on the importance of private initiative and individual entrepreneurship, d) focusing on the importance of specialization of areas and sector segmentation through comparative advantages e) focusing on the importance of the sector movement from secondary to tertiary activities, from manufacturing to services, f) exposure to the power of national and international markets, and g) penetration into other markets (Hall, 1998:426).

The tourist industry is recognized as a significant employment generator in Slovenia (Sorm&Terrell, 2000:450; Wanhill, 2000:135), as well as an important factor in the transition from socialist to capitalist forms of income (Szivas&Riley, 1999:460). Sorm and Terrell's studies in the Czech Republic have shown that - in terms of employment dynamics - the most dynamic sectors that have significantly improved the quality of transition, are the tourist industry, construction industry, financial services, retail and wholesale trade (Sorm&Terrell, 2000:440).

The tourist industry plays a special role in that process because of its involvement in many economic sectors and, as a result, its ability to mediate and correlate among the various sectors, in particular (Fletcher&Cooper, 1996:188).

In many restructuring processes from socialist to capitalist economies the tourist industry has been seen as one of the key economic branches able to contribute significantly to that process (Koscak, 1998:83, Szivas&Riley 1999:749, Fletcher&Cooper, 1996:182, Clarke et al., 2001:199, Vukovič, 2001:279). In the process of unification in Germany tourism was considered as an economic branch that would promote the convergence of the economies by initiating joint economic development (Coles, 2003: 199).

Tourism is also viewed as the economic branch that plays an important role in regional cooperation (Wanhill, 2000: 152). This importance and involvement of tourism in general economic and social streams is not only an issue relating to the development of one economic branch, but to the social development of each country that became an equal member in the EU in January 2004.

3. Goal and the Research Subject

Our research focused on the attitudes of Slovenian travel agency managers, and our goal was to determine the nature of these managers' attitudes to the status their agency would have as a result of Slovenia's admission into the European Union. We also wanted to know whether or not their agency had already been preparing for that new situation.

4. Sample

We were interested in travel agencies. Managers of 54 travel agencies in Slovenia, which is one third of all Slovenian travel agencies (their exact number was 161 at the time of our research), answered our questions.

36 or 66.7% of the respondents were men, 14 or 26.9% were women, and 4 agencies forgot to mark their gender. The fact that there are much more men in leading positions in the travel agencies is rather surprising, considering the strong presence of women in all forms of social life in Slovenia, as well as in the economy.

30 or 55.6% of the respondents are managers and 21 or 38.9% are operatives at their agencies. Interestingly, those respondents working for older agencies that have existed for a longer period of time are mostly operatives, whereas the respondents in younger companies were managers.

The respondents are between 23 and 58 years old. We grouped them in four categories based on their age, which can be seen in Chart 1. The average age of the respondents is 34.8 years, the median value is 34 years, and the mode is 26, so that there are more young managers than older ones. That the managers are younger people, the importance of the «gerontocracy» approach that was popular during the socialist period is declining, and the number of younger people taking over functions that demand dynamics and mental agility is increasing, seems to be part of a general process in Slovenia. Those respondents from older companies were mostly older and those from younger companies were younger in age. For example, in the case of companies that have been in business for over 11 years (75% of our sample), the respondents are mostly from the oldest age group, that is 56.3% of the respondents are over 42 years old.

The respondents have vastly differing work experience ranging from just one year, to one respondent with 37 years of service. The average is 11.7 and the median is 11 years of service. This means that half of the respondents had up to 11 years of service at the time of the questioning and the other half over 11 years of service. However, since the mode is 4, i.e. the most frequent value is 4 years of service, we can say that our group has rather minimal experience in the tourist industry.

Table 1: Respondents by age

Age group	Number	%
Up to 26	14	25.9
27-34	11	20.4
35-41.5	12	22.2
Over 42	12	22.2
No details	5	9.3
Total	54	100.00

Source: analysis of questionnaires

When it comes to working at their current agency, the range is also from one to 37 years of service. The average is 8.5 years of service at the current agency, the median is 6 years (half of them have been working there up to six years, half of them over six) and the mode here is also 4 years, i.e. most of the respondents had 4 years of service at the agency they were working at during the questioning.

Table 2: Respondents' education

Secondary	20	37.04
Two-year	14	25.93
University	17	31.48
No details	3	5.56
Total	54	100.00

Source: analysis of questionnaires

As we can see, most of our respondents are educated to secondary level but there are nearly as many respondents with a university education. The fact that there is a higher frequency of university education among operatives than among managers is very interesting. It tells us that the level of education is not the crucial criterion when it comes to appointing leading managerial functions and that there are probably other criteria that are more important than the level of education, while on the other hand, high levels of education are judged to be necessary for operative positions. The fact that operative positions in tourism demand a knowledge of foreign languages, which are best learned at universities, might be important. 52% of the operative respondents have a university education, whereas only 19% of the operatives were educated to secondary level. The situation is reversed among leading personnel. Only 20% of them have a university education, and 53% secondary training.

5.   The Research Tool

Our tool was a questionnaire with the basic purpose of investigating the networking of travel agencies in Slovenia, so that the subject matter of this article was only a minor part of our research tool. The tool had a total of 38 different types of questions. 13 of the questions had a Likert scale, 8 were yes/no questions, it was possible to give several different, although structured answers to the next 8 questions, and the last 8, that were mostly about population, were completely open (age, activities of the company, year of foundation, number of employees, etc.).

The questions we used in our research were either dichotomous or contained a Likert scale from 1 to 5, in which 1 represented total disagreement with the given answer, and 5 complete agreement.

6.   Methods

We mailed the questionnaire to all travel agencies in Slovenia; there were 161 at the time. 54 travel agencies answered on time, which means that the level of the received answers was 33.54%, a relatively high number of returned questionnaires for a mail research.

We processed the data using the software package SPSS, employing the distribution of frequencies, cross tabulation and the t-test for data processing.

7.   Results and Interpretation

As we had already mentioned in the introduction to the article, the basic goal of the research was to analyze travel agencies' attitudes towards Slovenia's admission into the EU, more precisely, to determine if the admission was causing fear or discomfort in travel agencies, given that the tourist industry of a small country would suddenly find itself exposed to strong competition from tourist agencies of countries with a much longer tourist tradition and a far better developed tourist market.

We tried to «cover» four variables with our questions. The first question was a claim that there is a possibility of doing business without interference, the same as before admission into the EU, in the second question we tried to find out if the respondents thought that there was strong competition in the EU and that there would be some difficulties in cornering the market because of that. We expressed our concern over the market clash with international companies in question number three. The fourth question was whether or not the agency had already been preparing for this important change. As we had already mentioned in the introduction to the article the measuring scale was a Likert scale, in which 1 represented total disagreement with the given answer and 5 complete agreement.

In diplomatic terms, we can say that these answers reflect moderate optimism or at least optimism taken with a pinch of salt. Our respondents agree with the claim that their agencies will freely continue to do business and they have been intensively preparing for the change but they don't know for sure if they will encounter difficulties in the market due to stronger competition and they disagree with the claim that they are afraid of conflict with international companies.

Table 3: Managers' opinions about their agency's position after Slovenia's admission into the EU

Attitude to their agency's position after Slovenia's admission into the EU	Mean	SD
1 We will freely continue to do business	3.51	1.38
2 We will have difficulties cornering the market due to stronger competition	3.00	1.40
3 We are afraid of the battle with foreign companies	2.02	.97
4 We have been intensively preparing for the change	3.04	1.29

Source: analysis of questionnaires

We tried to analyze these claims by controlling some of the variables describing certain characteristics of our respondents.

We tried to determine if the type of travel agencies - both those respondents whose major activity is the sale of services and those that operate tours - affected the attitudes to the claims. There is no statistically significant difference in their attitude to the first three questions. However, there is a statistically significant difference when it comes to the intensive preparations for the change. Managers at tour operator agencies claim that they have been preparing for the change (average grade is 3.45 with a standard deviation of 1.30), salesmen of tourist services say that they actually haven't been preparing for the change (average grade is 2.55 with standard deviation of 1.14). The t-test value is 2.46 and p=.018. The most probable reason for the difference we see above is the fact that tour operators have always been in some form of indirect contact with the Western European tourist market that they planned to become full partners of in January 2004, so that their preparation was a standard approach to treating the market they already knew well.

We also analyzed the relations between the claims among the already mentioned groups of agencies. We found a high but negative correlation of their attitudes to the first and second question both for tour operators and for salesmen (tour operators have Pearson correlation coefficient of -.80 and it is significant at .001, and salesmen have a correlation coefficient of -.84 and it is also significant at .001). This means that in both groups those who think there have been no negative changes at the same time feel that they will face no difficulties from stronger competition in the market. This is where the similarities in the interior correlations end. There is no more statistically significant correlation between the questions among the tour operators. However, there is a high and statistically significant negative correlation between the first and third claim among the salesmen (-.796 with significance of .001), which means that those who think they will freely continue to do business are not afraid of any conflict with foreign companies. At the same time those salesmen who think there have been no problems with cornering the market (arithmetic mean of the salesmen's answers to question number 2 is 2.85 with a standard deviation of 1.31, which means that they don't think they will have problems in cornering the market because of the stronger competition) are not afraid of any conflict with foreign companies (arithmetic mean for salesmen is 2.00 with a standard deviation of 0.86). The correlation between these two answers is .750 (in statistics that is a criterion for high correlation) and it is statistically significant at .001. Interestingly, the situation with the tour operators is different. First, there is no statistically significant correlation between the answers to these two claims. At tour operators the arithmetic mean for question 2 is 3.14 with a standard deviation of 1.49, which tells us that, unlike the salesmen, they are a bit more reserved and think that stronger competition is behind a fear of being able to corner the market. At the same time, they are not afraid of that conflict (an arithmetic mean for question 3. is 2.09 with a standard deviation of 1.07).

This could point to the likely explanation of the statistically significant difference in the answer to question 4, in which the tour operators, unlike the salesmen, pointed out that they had been preparing for the changes. The cause of their preparations is probably a greater fear of being able to corner the market in the wake of stronger competition in the EU. This probably doesn't say much about the differences in the type of activities between the tour operators and services salesmen, more likely it is about the different status they have in the tourist market. However, we believe that the variance is partly due to the age differences, which is related to different life experiences and, as a result, to the assessment of the economic effectiveness of the company.

Specifically, when we control the age variable we find interesting statistical differences among the groups. There are differences in questions one and two and they are statistically significant.

We grouped our respondents roughly into two age groups, one for those who were up to 34 years old at the time of questioning and one for all respondents over 34. We ended up, roughly, with two age groups that we constituted by the median, which was exactly 34 years. We thought that a more precise categorization by age was unnecessary. In the first question the younger group barely achieved a 3, i.e. its arithmetic mean is 3.04 with a standard deviation of 1.49, and the arithmetic mean of the other group is 4.05 with a standard deviation of 1.123. This tells us that statistically significant optimism is expressed more by the age group of older respondents than by the younger respondents (the t-test value is -2.47 with significance of .017). The situation is reversed in the second question, which was expected, of course, and for that question the arithmetic mean of the younger group is 3.36 with a standard deviation of 1.38 and the arithmetic mean of the older group is 2.35 with a standard deviation of 1.37 which points out that the older respondents are less afraid of the competition (the t-test value is 2.33 with significance of .025). The older group is more optimistic at this point as well.

It should be mentioned that the tour operators and the salesmen are in a similar situation when it comes to the age of the respondents (χ^2 test between the age groups, with 1 as a freedom degree is 0.20, which is far below the critical significance value of the χ^2 test with 1 freedom degree). So, if age is not the defining characteristic for the difference between the tour operators and the salesmen, then we can conclude that the differences we discovered in the attitudes of these two age groups are probably due to general life experience, rather than professional orientation as a tour operator or a salesman.

We were interested in two more variables relating to the attitudes towards the changes in the agencies that are expected to occur in the wake of Slovenia's admission into the EU: a variable of membership at one of the reservation networks and a variable of previous connections with other agencies.

When it comes to the previous connections with other agencies, then 40 respondents or 74.1 % claim that their companies have had previous connections with other companies and 13 respondents or 24.1% say that their company has never had any connections with other companies. There is a significant statistical difference in the first two questions. Respondents from the companies with previous connections have an arithmetic mean of 3.76 with a standard deviation of 1.20 for that question, which shows that they think their company will freely continue to do business, whereas respondents from the companies with no previous connections have an arithmetic mean of 2.77 with a standard deviation of 1.59 for that question and believe that their companies will not continue to do business without interference. The difference is statistically significant at a level of 0.03, and a t-test value of 2.29. The situation is similar with the answers to the second question. Managers of the companies with previous connections disagree with the claim that it has been more difficult to meet the market standards because of strong competition (the arithmetic mean is 2.65 with a standard deviation of 1.28), and those from the companies with no previous connections agree with the claim that it has been more difficult to corner the market because of the strong competition (the arithmetic mean is 3.77 with a standard deviation of 1.36). The t- test value is -2.61 with significance of 0.01. It seems that the managers from the companies with no previous connections with other companies fear for the success of their company once Slovenia joins the EU. Paradoxically, neither is afraid of the competition with foreign companies and neither have been preparing for the conditions they will face after admission.

There are no statistically significant differences in the four questions about the prospects of the company following admission into the EU in terms of participation in some reservation networks. It means that the management's attitudes towards the prospects of the company do not differ, regardless of their membership at reservation networks. In contrast to the large number of respondents (over 70%) who maintain other connections as addressed in the previous question, scarcely 63% of the respondents are members of a reservation network.

In the end, we have also analyzed whether the gender of our managers affected their attitudes to the questions we discussed. The t-test results found that there were no significant statistical differences whatsoever, that is to say both female and male respondents thought the same or very similar when it came to the claims we discussed.

8.   Conclusion

Slovenia's admission into the EU will definitely be a significant social step in many areas. Although the country itself has been preparing for many years, and adjusting its legislature, it will take longer for the residents' to adjust their attitudes and opinions.

When it comes to the tourist industry and its managers, it seems that there are some differences in assessing the prospects for the tourist companies after admission. It should be stressed that our research results show that the managers are aware of that, and that some of them have been preparing for the changes. However, the tour operators have been preparing more intensely. They have a much more complex role due to the nature of their business and their part in tourism, so they probably have a much broader view of the problem and a better understanding of the effects that such a big change will bring next year. According to our research, status within and experience of the market play the most important role in promoting the awareness of the meaning and importance of the change, not years of service and managers' life experience. Our research should definitely be continued and expanded, given the importance of the subject matter.

References

Baláz, V. (1998): Japanese tourists in transition countries of CE: present behaviour and future trends, *Tourism Management*, 19(5):433-443.

Brunn, S., Cottle, C. (1997): Small States and Cyberboosterism, *The Geographical Review*, 87(2):240-258.

Clarke, J., Denman, R., Hickman, G., Slovak, J. (2001): Rural tourism in Roznava Okres: a Slovak case study, *Tourism Management*, 22:193-202

Coles, T. (2003): The Emergent Tourism Industry in Eastern Germany a Decade after Unification, *Tourism Management*, (article in press)

Fletcher J., Cooper, C. (1996): Tourism Strategy Planning, Szolnok County, Hungary, *Annals of Tourism research*, 23(1):181-200.

Gosar L., Roblek I., (2001): Development problems in Slovene rural areas, *Mitteilungen der Oestereichischen Geographischen Gesellschaft*, 143:131-148.

Hall, D. (1998): Tourism development and sustainability issues in Central and South-eastern Europe, *Tourism Management*,19(5)::423-431.

Koscak M. (1998): Integral development of rural areas, tourism and village renovation, Trebnje, Slovenia, *Tourism Management* 19(1):81-85.

Langlois, S. M., Theodore, J., Ineson, E. M. (1999): Poland: in-bound tourism from the UK, *Tourism Management*, 20:461-469.

Meglič, J., Šmitek, B., Vukovič, G. (2005): Epromotion for a tourist destination. *Organizacija* (Kranj), 38(3):137-148

Pesci, EESC Rapporteur (2003): *Developing the Slovenian tourist industry – possible models*, 3rd meeting of the EU-Slovenia Joint Consultative Committee on the 12-13 June 2003 in Portorož, Slovenia.

Sorm, V., Terrell K. (2000): Sectoral Restructuring and Labor Mobility: A Comparative Look at the Czech Republic, *Journal of Comparative Economics*, 28:431-455.

Statistični urad Republike Slovenije, Ljubljana, 2002.

Szivas, E., Riley, M. (1999): Tourism Employment During Economic Transition, *Annals of Tourism Research*, 26(4):747-771.

Vukovič, G. (2001): Strategical management, marketing and strategy in non-profit organisations. V: *Znalostný manažment - kl'úč k úspechu* : proceedings. Bratislava: Dom techniky ZSVTS, 270-281

Wanhill, S. (2000): Small and Medium Tourism Enterprises, *Annals of Tourism Research*, 27(1):132-147.

Williams A. M., Baláz, V. (2002): The Czech and Slovak Republics: conceptual analysis of tourism in transition, *Tourism Management*, 23:37-45.

Milan Ambrož

"A Third Way" of Tourism Planning: Case of Slovenia

Abstract

Tourism is becoming a global industry involving hundreds of millions of people traveling both internationally and within their native countries each year. It is an important manifestation of modern times, and an interdisciplinary approach is needed to understand how this development can contribute to a positive development of the global world because it often plays a major part in the economy of poor countries and a very important role in more developed economies. According to empirical findings, there is a strong need to develop soft and more sustainable tourism, which should bring the greatest possible benefit to all participants – tourists, the local host population and the tourist businesses.

The sustainable approach to tourism development proposed by Burns (2004) named "A Third Way", integrated with Mlinar's (2001) early "accessibility" concept of such development was tested in our study. Our research shows that the active involvement of all stakeholders in tourism planning and decision-making at the levels of government, local community, business and local residents, is the key to developing sustainable tourism. The study confirms that both Slovenian tourism development documents and the law support an active role of stakeholders in the tourism development planning and decision-making process. On the other hand they show a strong tendency to control and somehow suppress the entrepreneurial initiative at the level of local community that could influence the future development of tourism in Slovenia.

Key words: sustainable tourism, planning, accessibility, a third way, reduced exclusion

1.   Introduction

Burns (2004:24) argues that a conventional approach to planning (Butler, 1980:9; Gunn, 1994:9) is very successful for developing mass tourism but it is highly probable that it will fail where it does not fit in the local environment. Pearce *et al* (1996:1) propose a more long-term and more sophisticated planning of tourism that will include all social agents at all societal levels. They emphasize the community reactions to tourism, which can be the basis for tourism planning, tourism and economic« growth at regional and local level and project development in the global environment. Pearce *et al* (1996:2) lead the attention of tourism planners to the importance of information and knowledge about tourism that encompasses the understanding and thoughts of the communities. Information and knowledge foundation is a vantage point for social action of the communities to plan and build sustainable tourism that is "soft and humane" as Krippendorf (1987:106) used to say.

The goal of future development of tourism in the global world is to create "resident-responsive tourism" (Ritchie, 1993:203). Sustainability in tourism is evidenced by the transition of the time span paradigm away from short term to long term planning and "destination visioning" (Ritchie, 1999:273). Cooper (2002:4) defines destination visioning as a community-based strategic planning approach that effectively places the future of the destination in the governance of the local community, government and industry. For him, the community-based strategic planning approach is the most important element of the tourism system. Its role is to motivate, deliver visitor experiences and to contribute to enduring memories of the tourism experience. Further, he develops the model and argues that the »increased growth of demand for tourism, coupled with the changing nature of the tourism consumer means that destinations are under pressure to be both competitive and sustainable«.

Mlinar (2001:779) thinks that these tendencies can be observed as a paradox between the preservation of the local distinctions through natural and cultural heritage and as a »marketing of identity« at the same time. This paradox can be solved only with the active role of the local community in all tourism development processes.

2.   Social Agency and Tourism Development

Social action introduced by Max Weber (1947:115) is the starting point of many social development politics. Weber understands it as an expected rational goal, as a tradition, and as a rational-emotional action, that results in a certain emotional state. Even managers in the field of tourism (WTO, 1997:16) recommend it to all countries as a strategy for inclusion of local communities in the process of planning, executing, controlling and evaluating the tourism politics, programs and projects at all societal levels.

Butler (1997:121) analyzed literature about tourism and discovered that there is little emphasis on the development of tourist destinations. This is one of the reasons for the limited number of theoretical concepts in this field. The second reason lies in the fact that development of tourism at global level and theoretical concepts cannot follow this development. Ritchie (1993:214) opines that fundamental changes in societal structure and changes in people's life style are caused by technological innovations and tourism development. Some research studies deal with the role of local community and support her thesis of the development of tourist destinations (Krippendorf, 1987:132; Hawkins, 1994:266; Ritchie, 1993:277; Murphy, 1981:195; Pearce in Moscardo, 2002:40; Mbaiwa, 2005:203).

Simpson (2001:6) advocates a similar view when he understands the high quality of tourist experiences, and good tourist and host relationships founded on a principle of social equality and inclusion of the local community in all social networks, as a successful development tourism strategy. Middleton and Hawkins (1998:16) support the harmonic development of tourism through social, cultural, physical and political features of the tourist destination. Gursoy et al (2002:79) emphasize the motives of local residents to participate in the development process, their capabilities to employ resources for tourism development. They argue that successful tourism development planning depends on two factors. The first one is the condition of the local economy and the second is the concern of local community for tourism development (Gursoy et al, 2002:105).

Hawkins (1994:261) lists nineteen points of tourism's social impact on the local community. Ritchie (1993:275) advocates the initiatives in the process of tourism development, and Murphy (1981:190) sees the perceptions of local residents as a base for defining priorities in tourism development. Krippendorf (1987:16) builds on the participating role of the local residents in tourism planning, and Andriotis (2002a:59) would integrate tourist structures, local residents and tourists and set the integrative role of local community as a key success factor in tourism planning. Vaughan, Jolley and Mehrer (1999:118) see the integrative, strategic, and advertising role of local community in tourism development.

The study of entrepreneurial potential in eight Slovenian local communities confirms that there is a positive relationship between community identity, its historical, cultural and political realities, and the entrepreneurial behavior of the individuals. The main drivers of the entrepreneurial action that are emphasized in the study are: (1) the conditions that enable the exploitation of the new entrepreneurial ideas (room for individual initiative), (2) tourism service quality, (3) sustainable planning, and (4) environment protection and safety. Study also shows that future development of entrepreneurial initiative in tourism in the local community should be based on proper destination management, open and democratic social environment, vision and strategy of the future and the conditions for individual entrepreneur agency (Ambrož and Kribel, 2002:13).

We conducted the second study in 26 local tourist organizations. The study targeted the development strategy of the Slovenian local communities in the field of tourism development. It revealed some very important flaws in the local tourism strategy such as the lack of entrepreneurial initiative and the shortage of quality information about the development opportunities. Only a few stakeholders are involved in tourism planning in the local community. There are not many internationally competitive tourism products and the participation of the Slovenian local communities in international tourism networks is weak (Ambrož, 2004:5).

3.   "Zero Sum Game" in Tourism Development

Globalization is a process that imposes change on social structures at local, national and global level. Mlinar (2001:767) uses three concepts to explain the distribution of power at these levels that is triggered by global changes. The first one explains the distribution of social power by "zero sum game" which is based on the mutual exclusion of societal levels, the second is the sharing principle, which empowers all societal levels, and the third distributes power through polarization and convergence. Mlinar (2001:767) critically revises these concepts and advocates the panacea of "globalization" based on the concept that

> "All societal levels are somehow intertwined and local social agents are present all over the world. At the same time, the world is more and more involved in the local unit and represents a kind of "microcosmos" (Mlinar, 2001:767).

This concept is a good starting point for explaining the role of tourism development at local community level. There is no doubt that tourism is a unique manifestation of modern times. It is a very complex activity of a global nature realized at local level. Urry (2001:215) goes beyond the complexity and argues that tourism is a complex and contradictory process. This is the reason why he is not surprised that there has been much discussion about the desirability of tourism as a strategy for economic development.

On the other hand, Mason (2003:3) is convinced that tourism is becoming a global industry involving hundreds of millions of people in international as well as domestic travel each year. Economic globalization processes in the field of tourism continue to exert pressure on mass tourism development. The effects of mass tourism rest on the mutual exclusion of the global and local societal levels. Krippendorf shows this exclusion when he evaluates the alternatives to mass tourism, the effects of which he contends can often be "devastating, inhuman and self-destructive" (1987:106). Krippendorf (1987:106) is critical of contemporary mass tourism, but he is convinced that these practices can change for the sake of both tourist and the host community. He advocates the panacea of a "soft and humane tourism" for which

> "The common goal must be to develop and promote new forms of tourism, which will bring the greatest possible benefit to all participants – travelers, the host population and the tourist business, without causing intolerable ecological and social damage" (Krippendorf, 1987:106).

Urry (2001:194) agrees with Krippendorf when he argues that the costs of congestion and overcrowding are immense because there is a strong urge to satisfy growing needs for authenticity, adventure, sensations and spiritual experiences. Urry (2001:215) adds that costs are high, but benefits from tourism are often less than anticipated because transnational companies invest in tourist destinations in the developing countries and in countries in transition and retain the majority of tourist expenditure. Mbaiwa (2005:219) confirms Urry's statement with his findings. He discovered that local economies often receive only a small return of the tourism benefits but have to bear the environmental costs caused by foreign companies. He suggests that for sustainable tourism development to be achieved, the distribution of benefits and local participation and investment in the tourism sector is very important.

Government officials, managers and entrepreneurs firmly believe in the economic potentials of tourism. They also believe that tourism is a driver of economic growth in areas that are severely damaged by the decline of local industry. Yunis (2004:12) analyzed the economies of some developing countries and discovered that they earned US$ 142,306 million in the field of tourism. In addition to this finding, Yunis discovered that tourism is the principal export in a third of all developing countries.

Nevertheless, Sinclair (1998:1) is skeptical of the positive impacts of tourism, because tourism involves considerable costs. Tourism boosts investments in infrastructure in the form of additional roads, airports, water, sanitation and energy. This expenditure is strongly related to tourism consumption rather than to common purposes. Making tourism a business involves a large amount of physical resources along with the investment in human capital. Sinclair (1998:49) sees the production of significant volumes of waste and changes in the social structure of the host community as the major problems of tourism development.

This problem was analyzed by Fadeeva (2000:6) who expects that the host population will be disturbed and the future of the tourist destination will be jeopardized if the social and economic capacity of the local community is not enough to withstand the pressure from tourism. These beliefs are not new as they are often shared by representatives of civil society, who are sometimes strongly convinced that tourism is a great threat to national culture, local identity, local resources and the environment.

There are clearly bipolarities that affect the distribution of power among social actors in tourism that is the issue in the first concept proposed by Mlinar (2001:767). We can observe the bipolarities in the relationship between the goals of transnational corporations that push mass tourism and are interested only in maximizing the tourism profit, and the development goals of local community based on sustainable tourism, which involves the inclusion of civil society and the controlled expenditure of the local resources.

As we already mentioned, tourism has its drawbacks and benefits for the whole society and especially for local communities. Despite many doubts about the positive effects of tourism, we must not forget that tourism is the only means of economic survival in many countries all over the world. Tourism is the only solution to dealing with unemployment in areas where conventional industry is in decline (Ambrož, 2004:5). The solution to this problem is not an expansion of mass tourism.

4.   "Third Way" Approach to Tourism Development

Burns (2004:24) has a slightly different view of tourism development when he identifies the failure of underdeveloped civil society to cope with economic progress and safety in the tourism areas in many countries. The result of such a condition is the inability to meet the expectations of the desired development. Burns (2004:24) indirectly supports an uneven distribution of power among transnational companies and the local community.

Burns (2004:26) is trying to equalize the distribution of power by seeking a compromise between the fulfillment of economic goals and the development expectations of the local community in the field of tourism. He adopts the "Third Way" approach developed by Giddens (1998:26), as the most promising concept, which includes the role of civil society in the relationship between the individual and the state. His decision is logical because the concept called "Third Way" is a way to help countries in transition to build social democracy to cope with changes in the emerging global world.

Burns' (2004:32) builds his conceptual framework on bipolarities that reflect the impacts of tourism, democratic development and tourism planning. The core question of "the Third Way" in his framework is *"who benefits from tourism development?"* (Burns, 2004:26). The first bipolarity is *holistic* in nature and reflects the concept of the universal development of tourism at the local community level, and the other is *economistic*, and reflects the idea of tourism development as a profit goal. The holistic bipolarity advocates independent and differentiated destination development with minimal dependency on the core. It aims at sustainable human development. The economistic bipolarity is aiming at maximizing marketing spread through known products.

As Burns (2004:27) states, there is a danger that the proposed model is oversimplified. In real life, the tourism planning process is full of contradictions and conflicts that are the result of frictions between *entrepreneur-led tourism* and *community-generated tourism* (Pearce *et al.*, 1996:2). Burns is convinced that this could be eliminated when the contradictions and tensions of the sustainable development are properly understood.

5. "Reduced Exclusion" and Local Community Involvement

When we look at these tensions and contradictions arising from the distribution of power among the social actors in tourism development, we must point out the contradiction resulting from the interplay between the interests of civil society and local authorities and the state, which controls and regulates tourism development. Mlinar (2001:769) tried to bridge this local-global distribution of power by converging it to a *"reduced exclusion"* of local democracy. His concept rests on building the network organization, which spreads beyond the local community and enables wider *accessibility*. Local communities become more powerful through their inclusion in transnational networks and less dependable on other territorial hierarchies. Dependency of the local community is spread among a larger number of actors in the development process, and the community becomes less dependent on a single party (Mlinar, 2001:771). The effect of this change of dependability is that the local participation becomes substantially richer and the distribution of power more balanced, because global problems become part of a local concern. Bipolarities receive a somewhat arbitrary position and are contingent upon the condition in the local community. The framework proposed by Mlinar (2001:771) is the basis for the Burns' model. Itzkowitz (1996:) proposed a similar concept when he advocated the idea that

"The micro and macro levels develop along distinct time and space paths, and are more accurately described as autonomous and as such, the relationship between the two levels is necessarily contingent on the indefinite development of each" (Itzkowitz, 1996:250).

6.   Tourist Destination Strategy

Planning is a very complex activity, so multiple definitions of this concept need to be developed. Management of tourist destinations cannot operate in a vacuum. A plan is needed to define the directions and goals. This is a very important component along with knowing where we come from and where we are now and what assets are at our disposal (Stutely, 2002:13).

Mason (2003:20) identifies tourist destination as a subject of tourism planning and management, because at the destination tourists encounter and interact with the local community and the local environment. Interaction between tourists and local inhabitants impacts the local population, the environment and the tourists themselves. This impact can be beneficial to the local community when tourism activity is successfully planned and managed. Development of the tourist destination is a very complex task because destinations develop and change over time, very different types of visitors at different times come to the destination and the involvement of local people in tourist destinations changes too (Ambrož, 2005:140).

Butler (1980:10) developed a model that suggests that tourist destinations change over time and that there are a number of linked stages: exploration, involvement, development, and consolidation. After the "consolidation" stage is in place, new possibilities emerge. The destination could "stagnate" without warning at a positive or negative level, it could "decline" or it could "rejuvenate". Agarwal (1997:70) tested Butler's development model in the study of Torbay in England. He found differences between the model and the application. The stages of development in Torbay were not discrete, there was very early involvement of locals in tourism to provide tourist infrastructure, the type of visitors to some parts of the tourist destination did not change over time. To avoid the decline of the destination, post-stagnation planning was enforced (Mason, 2003:25). Butler (1997:109) reacted to the criticism and pointed out that the key concepts of the model are dynamism, process and limits of growth. He emphasized management of the destination as a key factor to avoid the "decline" of the destination.

7. "Accessibility" of the Slovenian Local Communities

Slovenia is a country in transition with high unemployment problems caused by the low competitiveness of some industrial branches that are in constant decline. This is the reason for unemployment in some areas that have a large potential for tourist development. There are many dilemmas in terms of how to develop tourism in these areas and how to select the right development approach. Developed tourism strategies in Slovenia are entrepreneur-led in most developed tourist destinations.

Less developed regions in Slovenia do not have a distinct tourism development strategy yet. The new law on tourism development empowers local communities to build networks with partners from the public and private sectors and to cooperate freely on the base of common interests (Petrin, 2004:5). According to the economic policy of the Slovenian government, the development of tourism in Slovenia is priority and represents a public interest. Its development is based on the principles of sustainable tourism development and on the inclusion of all social actors in both Slovenia and the international environment.

We present the empirical research of our theoretical concept in this paper based on Mlinar's (2001:772) concept of "reduced exclusion" that we've converted to a tourism development model adding the bipolarities from the Burns' model "A Third way" of tourism planning. We empirically tested the proposed model to identify the potentials for the development of local tourism planning in Slovenian local communities. The tourism-planning model consists of five steps of local community tourism development.

The initial step (1) in the model is to analyze present tourism planning at the levels of the local community, region and national state.

The second step (2) rests on defining the benefits to all social actors involved in tourism planning. This is the critical step in planning and development because of the requisite consensus among all stakeholders concerning the development of tourism. This step could recur several times depending on the interest of the social actors in participating in the tourism development network.

The third step (3) delivers the answers to the questions of how to achieve the sustainable development, how to develop tourism as a system, how to develop a tourism service culture, how to build an open world system approach, and how to empower regional and local community tourism development.

The fourth step (4) in tourism planning develops the *global-local* approach that is based on the accessibility to the decision-making process and to the important social actors, on the participation of the weak partners in the planning process, on the empowerment of the opposition, and on the accessibility to the financial resources at an international level (Mlinar, 2001:773).

The fifth step (5) evaluates the benefits of tourism development and proposes the agreed tourism development strategy, which is contingent on the development conditions of the local community.

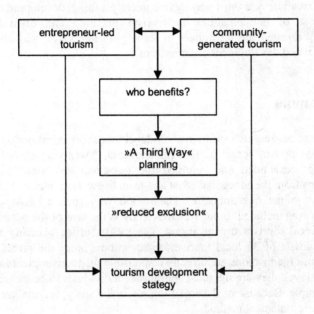

Picture 1: Model of "increased accessibility"

8.   Method

We tested the data in our study using the Wilcoxon matched pairs test for dependent samples because we gathered data from a small sample of respondents and assumed that the data was not normally distributed. This nonparametric test is alternative to the t-test for dependent (correlated) samples. The procedure of the test assumes that the variables under consideration were measured on a scale that allows the ranking of observations based on each variable (i.e. ordinal scale) and that allows the ranking of the differences between the variables. If the assumptions for the parametric *t*-test for dependent samples are met, then this test is almost as powerful as the *t*-test.

9.   Instrument

For the purpose of our research study, we constructed the instrument to test the proposed tourism development model, which consisted of a set of questions measured on seven and five point Likert scales. It consisted of three sets of questions that addressed third way development planning, development strategy, and the role of municipalities in tourism planning and decision-making processes. We tested the instrument in some pilot interviews with directors of local tourism and development organizations.

10.   Sample

We conducted our research study in all 36 local tourist organizations and in some regional development organizations in Slovenia. Twenty-seven local tourist organizations responded and returned the questionnaire. Nine local tourist organizations from the Slovenian coast and from the western part of Slovenia did not respond to the questionnaire. The reason why such a small sample of respondents was included in our research study is the size of the network of the Slovenian local tourism organizations. The local tourism planning system in Slovenia consists of 36 local tourism organizations. Since the directors of the local tourism organizations are directly responsible for tourism planning at local community level, they are the most appropriate individuals to be included in our research sample. Because of the small sample in the study, we analyzed it using nonparametric statistical methods.

11.   Results and Interpretation

Table 1 shows the findings from the analysis of attitudes towards "A Third Way" of tourism development model. Respondents gave priority to the primary role of the local community in the tourism development process (6, 03), to the system approach to tourism (5, 79), and to the development of sustainable human development (5, 54).

Table 1: "A Third Way" planning paradigm

Descriptive statistics			
"A Third Way"	Valid N	Mean	Standard Deviation
Sustainable human development	28	5,250000	1,578208
Tourism as a system	28	5,785714	1,286890
Tourism as a culture	28	5,535714	1,426785
Modern world system	28	4,964286	1,731669
Periphery	28	6,035714	1,290482
Tourism benefits	28	4,642857	1,909320
Universal development	28	4,214286	1,853068
Openness and local knowledge	28	4,535714	1,731669
Consensus seeking	28	4,571429	1,619981
Networking	28	5,428571	1,501322
Partnership	28	5,500000	1,710534
Individual agency	28	5,285714	1,606864
Horizontal participation	28	5,321429	1,564740
Inclusion in decision-making	28	4,392857	1,571489
Controlled private initiative	28	4,428571	1,730523

We matched the attitudes towards "A Third Way" of tourism in Table 2 with the attitudes towards the accessibility of the local community. Accessibility is the capability of the local community to build networks that spread beyond local and national levels, which results in richer participation and more balanced power distribution. We measured the accessibility on a power scale from one to five, where the lower number (1) indicates imposed tourism development forced by the government, and the higher number (5) signifies the active participation of the local community in all tourism development activities.

Those respondents that participated in the study see the accessibility of the local communities to the planning resources as a formal right to participate in the planning of tourism development and to share knowledge with all stakeholders in the planning process (average mean = 3, 93; stdev = 1).

The results show that the accessibility of the Slovenian local communities that participated in our research is strongly matched with periphery ($Z = 4, 2$), with the system approach to tourism development ($Z = 3, 97$), and with tourism as a culture ($Z = 3. 37$). The match with partnership building ($Z = 3, 16$), and with horizontal participation ($Z = 3, 11$), with individual agency ($Z = 3, 03$), with network building ($Z = 2, 99$), and with sustainable human development ($Z = 2, 82$) is somewhat weaker. The weakest match is with modern system building ($Z = 2, 70$). We can conclude from the results that there is a strong urge for legal and professional autonomy of the local community in the process of tourism development planning based on a system approach and long-term development.

Table 2: Matched "Reduced exclusion" and "A Third Way"

Wilcoxon Matched Pairs Tests Marked tests are significant at p <,05000				
"Reduced exclusion" x "A Third Way"	Valid N	T	Z	p-level
Sustainable human development	28	29,5000	2,818617	0,004823*
Tourism as a system	28	15,0000	3,968769	0,000072*
Tourism as a culture	28	32,0000	3,371429	0,000748*
Modern world system	28	32,5000	2,706619	0,006798*
Periphery	28	3,0000	4,200000	0,000027*
Networking	28	51,5000	2,986667	0,002821*
Partnership	28	51,0000	3,162048	0,001567*
Individual agency	28	56,0000	3,035058	0,002405*
Horizontal participation	28	26,0000	3,110810	0,001866*

* Statistical significance at 5 %

Table 3 reveals that the marketing spread (mean = 5, 85) is the most important strategy of tourism development along with tourism development obstacles identification (mean = 5. 48), agreed tourism goals (mean = 5, 40) and core tourist destination competences (mean = 5, 40). The findings reveal a product and goal oriented tourism strategy that is supported by the ongoing elimination of tourism development obstacles and the use of core destination competences.

Table 3: Tourism strategy

Descriptive statistics			
Tourism strategy	Valid N	Mean	Std.Dev.
Tourism potential	27	4,962963	1,697996
Sustainable tourism	27	5,037037	1,720498
Tourism development obstacles identification	27	5,481481	1,602171
Tourism goals consensus	27	5,407407	1,715523
Tourism structure and relationship development	27	5,222222	1,867399
Marketing spread	27	5,851852	1,433224
Tourism product diversification	27	5,111111	1,928198
Core tourist destination competences	27	5,407407	1,737798
Controlled private initiative	27	3,740741	2,141038

We matched "A Third Way" approach in Table 4 to community planning with the local community tourism strategy. The variables of periphery and controlled private initiative ($Z = 3, 74$), networking and controlled private initiative ($Z = 3, 47$), inclusion in decision-making and marketing spread ($Z = 3, 46$), tourism as a system and controlled private initiative ($Z = 3, 28$), common development and marketing spread ($Z = 3, 14$), have the highest match.

Controlled private initiative is the most expected strategy in the local communities in the study. It shows the system approach supported by open tourism development planning where a local community as the most important planner affects tourism development based on the initiative of the private companies. Common development and inclusion in decision-making processes enable a marketing spread of tourist products and services and are key variables for achieving the strategic tourism goals.

Table 4: Sustainable planning and tourism strategy

Wilcoxon Matched Pairs Test. Marked tests are significant at p <,05000				
"A Third Way" x Local community tourism strategy	Valid	T	Z	p-level
Sustainable system development & controlled private initiative	27	20,5000	2,650934	0,008027*
Tourism as a system & tourism potential	27	46,0000	1,971867	0,048626*
Tourism as a system & controlled private initiative	27	21,0000	3,284598	0,001021*
Tourism as a culture & controlled private initiative	27	28,0000	2,874616	0,004045*
Modern world system & marketing spread	27	35,0000	2,199293	0,027858*
Modern world system & controlled private initiative	27	28,5000	2,272229	0,023073*
Periphery & tourism potential	27	31,5000	2,743951	0,006071*
Periphery & sustainable tourism	27	44,0000	2,277293	0,022769*
Periphery & tourism structure and relationship development	27	23,0000	2,101459	0,035601*
Periphery & tourism product diversification	27	32,0000	2,106546	0,035158*
Periphery & controlled private initiative	27	15,0000	3,741039	0,000183*
Tourism benefits & development obstacles	27	36,0000	2,155743	0,031105*
Tourism benefits & marketing spread	27	29,5000	2,635863	0,008393*

Common development & tourism development obstacles	27	36,0000	2,575954	0,009997*
Common development & tourism goals consensus	27	48,5000	2,328763	0,019872*
Common development & tourism structure and relationship development	27	50,0000	2,053297	0,040045*
Common development & marketing spread	**27**	**25,0000**	**3,145567**	**0,001658***
Common development & tourism product diversification	27	57,5000	2,240130	0,025083*
Common development& core tourist destination competences	27	48,5000	2,722138	0,006486*
Local knowledge & marketing spread	27	17,0000	2,983199	0,002853*
Consensus seeking & tourism development obstacles	27	24,0000	2,857195	0,004274*
Consensus seeking & tourism goals consensus	27	26,0000	2,390574	0,016823*
Consensus seeking & marketing spread	27	34,5000	2,815370	0,004872*
Networking & tourism potential	27	19,5000	2,071624	0,038301*
Networking & controlled private initiative	**27**	**19,5000**	**3,473825**	**0,000513***
Partnership & controlled private initiative	27	47,0000	2,767761	0,005645*
Individual agency & marketing spread	27	23,0000	2,101459	0,035601*
Individual agency & controlled private initiative	27	36,0000	2,575954	0,009997*
Horizontal participation & controlled private initiative	27	44,0000	2,678416	0,007398*
Inclusion in decision-making & tourism development obstacles	27	47,5000	2,363520	0,018103*
Inclusion in decision-making & tourism goals consensus	27	30,0000	2,417045	0,015648*
Inclusion in decision-making & tourism structure and relationship development	27	47,0000	2,380899	0,017271*
Inclusion in decision-making & marketing spread	**27**	**9,0000**	**3,460828**	**0,000539***
Inclusion in decision-making & core tourist destination competences	27	35,0000	2,970607	0,002972*
Controlled capitalism & tourism development obstacles	27	68,0000	2,129047	0,033251*

Controlled capitalism & tourism goals consensus	27	59,0000	2,191431	0,028421*
Controlled capitalism & marketing spread	27	21,0000	2,977921	0,002902*
Controlled capitalism & core tourist destination competences	27	63,0000	2,281121	0,022542*

* Statistical significance at 5 %

12. Conclusions

The empirical investigation of our model based on the reduced exclusion of the local community in tourism development planning reveals that the community-generated paradigm of tourism development is preferred. Somehow these results show the cultural influence of a region that has been politically oriented towards the left since the end of WWII until recently. During this time, tourism in Slovenia was mainly a peripheral economic activity that was controlled and planned by the state. Nevertheless there are strong tendencies to change this relationship and the distribution of power in the tourism development process.

Respondents in our study are convinced that the local community should play the primary role in the development process. Tourism should be treated as a system with dynamic processes within the system and connected to the system environment. The long-term goal of planning at the level of local community should be sustainable human development.

When we matched »A Third Way« tourism planning model with the »reduced exclusion«, we discovered that formal conditions for the accessibility of Slovenian local communities are supported by law. Results emphasized the formal participation role of the local community in the planning process with the mandate to share knowledge with interested participants in the knowledge network. In addition, the results confirmed the theory proposed by Mlinar (2001:765) and Burns (2004:40) that increased accessibility is possible when the system approach to tourism planning is implemented and supported by cultural change.

The analysis of the formal tourism strategy of the Slovenian local communities shows that it is destination competitive as well as product and goal oriented and aimed at the ongoing elimination of obstacles to the marketing spread of tourism products and services. That kind of strategy is entrepreneur-led and demands open system planning.

When we matched it with the »Third Way« planning paradigm, we discovered that local communities show a strong tendency to control private initiative in tourism development. This tendency also shows that there is an interest for open tourism planning only if the boundaries of tourism development are set by the local community and implemented by the private companies. These results are somewhat in contradiction with the need for increased accessibility to balance the distribution of power in the global world. Common development as an ultimate goal is threatened by the expectations of the local community to play the role of the planner, which sets limits on the development and boundaries of accessibility.

We can conclude from the results of the study that the increased accessibility of the local community is the hierarchical approach to tourism development planning. It is obvious that the culture of risk avoidance at the level of local community should change to a culture of opportunities in the future to pave the way for more accessible and entrepreneur-led tourism development.

References

Agarwal, S. (1997): The resort life cycle and seaside tourism. Tourism Management 18 (2): 65-73.

Ambrož, M., Kribel, Z. (2002): Entrepreneurial initiative in the field of tourism in local community, *In*: Vadnjal, J. (ed.) *Global tourism and entrepreneurship in the CEI Region: Conference proceedings, 6 - 8 November 2002, Radenci, Slovenija.* Ljubljana: Small Business Development Centre: 7-17.

Ambrož, M. (2004): Local tourism development strategy. In: Ambrož, Milan (ed.). *Strategic partnerships for the development of tourist destinations: international conference: proceedings: mednarodna konferenca: zbornik prispevkov.* Portorož: Turistica, Visoka šola za turizem: 5.

Ambrož, M. (2005): *Sociology of Tourism.* Portorož, Turistica:140

Andriotis, K. (2002): Local Authorities in Crete and the Development of Tourism. *The Journal of Tourism Studies*, 13 (2):53-62.

Andriotis, K. (2002a): Resident's satisfaction or dissatisfaction with public sector governance. The Cretan case. *Tourism and Hospitality Research: The Surrey Quarterly Review*, 4 (1): 53-68.

Burns, P. M. (2004): Tourism Planning A Third Way ? *Annals of Tourism research*, Vol. 31, No. 1:23-43.

Butler, R. W. (1980): The Concept of a Tourism area cycle of evolution. Canadian Geographer, 24:5-12.

Butler, J. R., Wilson, D. C. (1990): Managing Voluntary and Nonprofit Organizations: Strategy and Structure, London, Routledge:16.

Butler, R. W. (1997): Modelling tourism development: Evolution, Growth and Decline, In: Wahab, S. and Pigram, J.J. (eds.) *Tourism Development and Growth - The Challenge of Sustainability*. London: Routledge:109-128.

Butler, R. W. (2002): Problems and Issues of Integrating Tourism Development. V Pearce, D.:22.

Cooper, C. (2002): Sustainability and tourism visions. VII Congreso Internacional del CLAD sobre la Reforma del Estado y de la Administración Pública, Lisboa, Portugal, 8-11 October. http://unpan1.un.org/intradoc/groups/public/documents/CLAD/clad004 4548.pdf, 09.06.2005.

Dann, G. M. S. (2002): Theoretical Issues for Tourism's Future Development. V Pearce, D. G., Butler, R. W. (ed.) *Contemporary Issues in Tourism Development*, Routledge Advances in Tourism: 13-31.

Fadeeva, Z. (2000): Small and Medium-Sized Tourism Enterprises in Sustainable Development Networks. *Greener Management International*: 6, 22.

Giddens, A. (1998): The Third Way. Polity Press.

Gunn, C. (1994): A perspective on the purpose and nature of tourism research methods, In Travel, Tourism, and Hospitality Research (2nd Ed.), J. R. B. Ritchie and C. R. Goeldner, (eds.). New York: John Wiley & Sons:3-11.

Gursoy, D., Jurowski, C., Uysal, M. (2002). Resident Attitudes: A Structural Modeling Approach. Annals of Tourism Research, 29 (1): 79-105.

Hawkins, D.E. (1994): Ecotourism: Opportunities for Developing Countries, In: W.F Theobald ed.: Global Tourism. The Next Decade; Butterworth-Heinemann, Oxford, UK:261-273.

Krippendorf, J. (1987): *The Holiday Makers: Understanding the Impact of Leisure and travel*. London: William Heinemann:16-106.

Mbaiwa, J. E. (2005): The Problems and Prospects of Sustainable Tourism Development in the Okavango Delta, Botswana Journal of sustainable tourism, Vol. 13, No. 3: 203-306.

Middleton, V. T. C., Hawkins, R. (1998): *Sustainable Tourism: A Marketing Perspective*. Harlow: Addison Wesley Longmann:16.

Mlinar, Z. (2001): The Empowerment and Weakening of Local actors and the emerging and fading of local features in the globalization process. Theory and praxis, vol. 35, no. 1: 765-785.

Murphy, P. E. (1981): Community attitudes on tourism: A comparative analysis. Tourism Management 2, 2: 189-195.

Pearce, P. L., Moscardo, G. (2002): Tourism Community Analysis. V Pearce, D. G., Butler, R. W. (ur.) *Contemporary Issues in Tourism Development*, Routledge Advances in Tourism: 31-51.

Petrin, T. (2004): Zakon o spodbujanju razvoja turizma. http://www.mgrs.si/ministrstvo/tema_meseca/index.php :5.

Ritchie, J. (1993): Tourism research. Policy and managerial priorities for the 1990s and beyond. V D. Pearce in R. Butler (eds.) *Tourism Research. Critiques and Challenges*, London, Routledge:201-216.

Ritchie, J. R. B. (1999): Crafting a Value-Driven Vision for a National Tourism Treasure *Tourism Management*, 20 (2):273-282.

Simpson, K. (2001): Strategic Planning and Community Involvement as Contributors to Sustainable Tourism Development. *Current Issues in Tourism*, 4 (1):6.

Sinclair, M. T. (1998): Tourism and economic development: A survey. *Journal of Development Studies* 34 (5):1–51.

Stutely, R. (2002): The definitive Business Plan. Second Edition, New York, Financial Times: Prentice Hall:13.

Urry, J. (2001): The Changing Economics of the Tourist Industry. In Apostolopoulos, Y, Leivadu S., Yiannakis, A (eds.) The Sociology of Tourism. Theoretical and empirical investigations. London, Routledge:194-215.

Vaughan, D. R., Jolley, A. Mehrer, P. (1999): Local authorities in England and Wales and the development of tourism Internet sites. Information technology and Tourism, 2 (2):115-129.

Weber, M. (1947): The Theory of Social and Economic Organization. Talcott Parsons (ed.), New York, The Free Press:115.

WTO (1997). *World Tourism Leaders' Meeting: The Social Impact of Tourism. Final Report*, Madrid, Spain: World Tourism Organization:16.

Yunis, E. (2004): *Sustainable Tourism and Poverty Alleviation*. World Bank – ABCDE Conference - Europe, 10 May, Brussels:12.

Gorazd Sedmak

Differentiation of Catering Outlets as a Variable in Tourist Destination Positioning

Abstract

A tourist destination is often seen as an integral tourism product consisting of lodging establishments, food and beverage operations, tourist attractions, animation programs, information, etc. Food, besides lodging, is the only indispensable element of an integral tourism product. Although food is rarely the most important motive for visiting a destination, tourists usually take the opportunity to taste and become acquainted with the local food of a region. Thus, local food can be claimed to be a natural and at the same time "obligatory" destination attraction element. Of course, not all tourists have the same affinity for the specific gastronomic products being offered. This article presents some findings and reflections derived from two studies that were conducted in the Slovenian coastal area. The discussion claims that differentiation in the supply of gastronomic products is an important base for tourist destination positioning.

Key words: product differentiation, supply of gastronomic products, positioning of tourist destinations, Slovenian coastal area.

1. Introduction

Since the late 1970s the appearance of new tourist destinations and a sharp drop in long-haul prices have caused serious problems for many traditional mass tourist destinations. Many of them have faced decreased economic viability due to declining profit margins and the reduction of tourist numbers and average spend per head. Besides, changes in holiday habits and underlying demographics have occurred (Agarwal, 2002:33; Willmott, Graham, 2001:31, Baum, 1996:27). In order to recover and adapt to the new conditions they have had to differentiate their products and cater only for a limited number of market segments (Priestley and Mundet, 1998:105, Carey et al., 1997:429). The process of destination repositioning by emphasizing the comparative competitive advantages is a comprehensive task that has to take account of all elements of the integral tourism product.

2. Tourist Destination and Food Supply

Tourist destination is a complex concept. It is usually linked to a geographic region with adequate tourism infrastructure, which is perceived and can be promoted as a limited entirety. Tourist destination is often seen as an integral tourism product, including lodging capacities, food and beverage operations, tourism attractions, animation programs, information, etc. (Rispoli, Tamma, 1995:26, Cooper et al., 1998:102). As Medlik and Burkart (1975:122) put it: " ... from the tourist's viewpoint the integral tourism product is the overall tourist's experience starting with the moment of leaving home and ending with his or her return home". Murphy, Pritchard and Smith (2000:44) set an interesting parallel between a tourist destination and a retail store. Different customers entering the store buy different products, however the ambience/"atmosphere", general price level, personnel attitude, etc. are the same for all of them. On the other hand, tourists buy a variety of products and services from different providers at a destination (lodging, meals, souvenirs, etc.), while the climate, the overall appearance and attractions of the destination are the same for all.

Food is a specific element of the tourist destination supply. On the one hand it represents a necessary condition for the existence of tourism at the destination, on the other hand, if we speak of typical local food, it can be seen as an important tourism attraction element and of course, an envoy of the cultural heritage of the local people (Quijano-Caballero, 2002:64, Kuznesof et al.:1997: 199). As such it can well be included in the process of positioning a tourist destination. Besides, if presented correctly it should constitute a sustainable competitive advantage for the destination. No one can imitate typicality without losing its entity.

3.   Segmentation and Positioning

It is generally accepted that no tourist destination can satisfy the needs of the entire population. As people differ in their daily lives, they also have different needs as tourists. It is hard to imagine a group of young extreme sports enthusiasts spending their holidays at the same resort as a group of elders. Therefore, segmentation and positioning of a tourist destination is needed. The conventional approach to segmentation is first to recognize the different reasons for the consumption of a product and then to distinguish the relative importance of the various product attributes for different consumers. This is followed by profiling the different groups of consumers on a demographic and lifestyle basis. Many scholars (Johns and Pine, 2002:3, Prentice et al., 1998:3, Gountas and Gountas, 2001:216) suggest that the personal values of tourists are a very valuable base for market segmentation in international tourism markets, as they predict the importance people attach to the various amenities at the tourist destination.

When considering the so-called "new tourists" and "cultural tourists", especially, one has to bear in mind that connoisseurship of foreign culture (including the local foods) gives tourists credit for intelligence and represents symbolic capital that is very important in the modern class struggle, as it forms the means to raise one's social position relative to other class fractions (Richards et al., 2001:74, Derret, 2001:16, Poon, 1998:122). No need to emphasize that price elasticity is much lower when a tourist sees a product primarily in this light. Undoubtedly, this is important when supplying gastronomic products.

After a target segment is defined, according to the integrated approach to market segmentation, positioning a tourist destination is the next step (Horvat, 2002:165). Competitive destinations, their positions, competitive strengths and weaknesses should be observed here. The positioning analysis is usually considered as a preparatory step towards new product planning or towards differentiating, or just improving the existing product (Dolnicar et al., 2000:40). It is simply all about deciding who we want to be on the market. Shall we cater for SSSS tourists, a narrow segment of special interest tourists or some other segment, or a bundle of segments?

4.   Differentiation

Differentiation as a competitive strategy often gives the product supplier the opportunity to set a premium price. This is only possible when the uniqueness of the product represents extra value to the buyer. This may be the case when it reduces the buyer's costs or when they get something more (or more convenient) at the same price (Porter, 1996:21, Holloway, 2004:132). This is especially important in the hospitality industry, for example in restaurants, where buying the product implies staying there for some time.

A product is differentiated if any basis exists to make it possible to distinguish it from similar products of other sellers (Pelham, 1997:178, Nickels, Wood, 1997:273). The basis can be real or imaginary. Theoretical conditions of differentiation are that buyers perceive products within a certain group of products as close substitutes, but as weak substitutes for products from outside the group. At the same time, products from within the group should not be perfect substitutes, i.e. every seller should deal with a descending demand curve. Although factors of differentiation may be intangible, buyers at least partially consider them in the selection process (Kotler, 1996:294). They virtually choose from among different products.

The image of a tourist destination is the collective sum of a tourist's beliefs, ideas, impressions, and expectations (Oliver, 2001:275). The consumption of destinations produces images that are based on the tangible as well as the intangible (i.e. atmosphere) components of these environments. These images influence tourists' perceptions of quality and value (Murphy at al., 2000:43). As we move away from the segment of mass tourists towards more demanding segments (cultural tourists, special interests tourists, eco-tourists etc.) the intangible components and symbolic values of destination elements become prevailing bases of tourism product differentiation.

The supply of gastronomic products in tourist destinations can be differentiated in several ways. Firstly, according to the type of food and beverage operation - this can be located within a lodging operation or can stand alone. Secondly, differentiation can be achieved through the price level, which should of course be commensurate with the given service quality level. Thirdly, the location of the operation is often a very important factor of differentiation. Undoubtedly, many more ways of how to differentiate the supply of gastronomic products could be found, nevertheless the contents of the supply is surely one of the crucial factors. If we keep it simple, the main issue is whether a tourist in a destination has the possibility to taste and become acquainted with the local food from the region, or is the supply of gastronomic products aimed primarily at satisfying the tourists' basic need for food.

5.   Research

The results of two different research studies are presented and discussed below.

The first project was carried out in 2000. The three Slovenian coastal municipalities were included in our research into the implicit prices of characteristics of catering outlets. All 115 outlets operating in this area were inspected and we managed to gather data for 102 of them. In order to enable a comparison of prices and average spending per visitor between catering outlets in tourist destinations and those located elsewhere, they were split into two groups. The first group comprised outlets located in tourist destinations, while the second group included the remaining catering outlets. We subsequently excluded 19 outlets whose ambiguous nature could create biased results as far as "the location in a tourist destination" is concerned. Hence, the final analysis covered a sample of 83 restaurants (Sedmak et al., 2004:160). Information on prices, average spending per person, products on offer and other characteristics were obtained for all of them.

The second research study was conducted in 2003 in the municipality of Piran - Slovenia's most important tourist destination. A random sample of 300 tourists was asked about the importance they place on local food as part of the integral tourism product in the destination choice process compared to the price level of the destination. The five point Likert scale was used. After excluding inadequate questionnaires, 274 underwent analysis.

6.   Results and Analysis

Some of the results of the first research study are presented in Table 1. Only 21.6 % of the catering outlets in tourist destinations had the house's typical dishes[1] on their menus in 2000, while 52.2 % of the outlets located away from a tourist destination offered such dishes. Outlets at tourist destinations had higher average prices as well as higher average spending per person due to the valuation function of tourism.

---

[1] Although in the research study we were not concerned whether typical house dishes were traditional local food. From informal discussions with owners we learned that this was the case for the majority of catering outlets.

Table 1: Comparison of restaurants in tourist destinations and restaurants located away from a tourist destination

	Location in a tourist destination	Location away from a tourist destination
N	37	46
Percent of outlets offering typical dish of the house	21,6	52,2
Price of a selected three-course meal in USD*	11,65 (St. dev. 1,57)	10,82 (St. dev. 1,20)
Average spending per visitor in USD*	11,69 (St. dev. 4,33)	8,48 (St. dev. 4,75)

*Values in USD. USD 1 = SIT 217.47 (exchange rate as at 31.5.2000, Bank of Slovenia, 2002).
Source: Sedmak et al., 2004

The main results of the second research study are presented in Table 2.

Table 2: Importance of the price level and availability of local food

Variable	Mean	St. dev
Importance of the price level	3,47	0,87
Importance of availability of local food in the catering outlets	3,80	0,89

Source: Sedmak, 2004

Results show that tourists in the leading Slovenian tourist destinations place higher value on the availability of local food than on the general price level in the destination choice process.

We conducted a K-means cluster analysis because the majority of demographic variables (education, gender, age) did not show a statistically significant impact on the allotted importance of the availability of local food. The results are presented in Table 3.

Table 3: Results of the cluster analysis

	Importance of availability of local food in the catering outlets	Education group	Age group	Income per family member group
Cluster				
1	4,11	2,86	4,33	7,80
2	3,56	2,20	2,56	5,63
Total:	3,81	2,51	3,39	6,64

Source: Sedmak, 2004

The algorithm produced two distinct clusters. The first cluster includes elder tourists with higher incomes and a higher degree of formal education. They value the availability of local food in catering outlets higher than members of the second cluster.

7.   Discussion

Although the presented results do not lead to straightforward conclusions, we will try to propose a few explanations that are, in our opinion, important for understanding the meaning of differentiation for catering outlets in tourist destination positioning. The Slovenian coastal area has some specific features, such as proximity to the Italian border, that undoubtedly make it unique. Thus, a great deal of independent restaurants and similar catering outlets outside the tourist destinations cater for daily visitors from Italy. These individual visitors are known for indulging in their love of good food but they are also well acquainted with the prices within and outside the tourist destination areas. This might be a reason why a high percentage of outlets (52,2 %) offer typical (local) food at relatively low prices. On the other hand, the tourist destinations we have been dealing with were typical mass tourism resorts in the past. Despite the changes in the tourist destinations' orientation towards wellness and MICE tourism that have taken place in the last decade, restaurant managers seem to stick to the type of supply that was tailored for mass tourists. As they nevertheless attain higher prices and average spending per guest than outlets outside tourist destinations, they do not see the need to change. It is our belief that the seemingly good results are possible only because of the valuation function of tourism, restricted mobility and the lack of information among the tourists. Seemingly, because if supply better matched demand, prices and average spending per guest could have been even higher. The results from Table 2 clearly show how high the expectations of the tourists are in terms of the supply of local food. However, many hospitality operation managers are not aware of the importance of this element and the potential it offers to their businesses.

The results presented in Table 3 support our statement, too. The cluster analysis also shows that differentiation in the supply of gastronomic products can be an important base for tourist destination positioning. Namely, both the empirical research and tourism theory suggest that the era of so-called "new tourists" is approaching. Unlike classical mass tourists, these tourists, tired of modernity and everyday role-playing, search for authentic experiences. On average, they are better educated and more willing to pay for the authentic tourism experience, part of which also includes typical local food. On the other hand, demographic trends show a constant ageing of the population in the world's main tourist generating countries. Seniors have become an important segment worth specializing in. Tourism theory suggests that elderly tourists are inclined to seek meaning from the past through nostalgia and thus prefer traditional/local to modern/global.

Of course the supply of gastronomic products in tourist destinations is only a part of the destination supply. A willful and sensible differentiation of the destination product and its positioning should adhere to the main rule - that is the consistency of the integral tourism product. This means that the mere introduction of typical foods on restaurants menus is not enough to ensure successful tourist destination repositioning, if not supported by high quality and carefully chosen entertainment programs, events showing the cultural heritage of the region, and of course adequate marketing activities. The changes that have been taking place on the Slovenian coastal area in the last decade show that hotel managers are aware of the trends and that there is willingness to keep pace with successful modern destinations. However, restaurant managers need to change their attitude and take a more active role in the process.

References

Agarwal, S. (2002): *Restructuring Seaside Tourism - The Resort Lifecycle.* Annales of Tourism Research, Vol. 29 1: 25-55.

Baum, T. (1996): *Images of tourism past and present,* International Journal of Contemporary Hospitality Management, Vol. 8, 4: 25-30.

Burkart, A. J., Medlik, S. (1975): *The Management of Tourism.* Heinemann, London: 122.

Carey, S., Gountas, Y., Gilbert, D. (1997): *Tour operators and destination sustainability,* Tourism Management, 18 (7): 425-431.

Cooper C., et al. (1998): *Tourism - Principles and Practice,* 2nd ed., Longman, New York: 102.

Derret, R. (2001): *Special interest tourism.* Douglas Norman, Douglas Ngaire, Derret Ros eds., Special interest tourism, John Wiley & Sons Australia, Brisbane:16.

Dolnicar S. et al. (2000): *A Tale of Three Cities: Perceptual Charting for Analyzing Destination Images.* Woodside ed al. ed.: Consumer Psychology of Tourism, Hospitality and Leisure, CABI Publishing, Oxon:40

Gountas, Y., J., Gountas, S. (2001): *A New Psychographic Segmentation Method Using Jungian MBTI Variables in the Tourism Industry,* in Mazanec et al. (eds.), Consumer Psychology of Tourism, Hospitality and Leisure, Vol. 2. CABI Publishing, Oxon:216.

Holloway, J.C. (2004): *Marketing for Tourism,* 4th ed., Prentice Hall, Harlow: 132.

Horvat, J., et al. (2002): *Segmenting Domestic Tourism Demand: Application of Factor and Cluster Analyses.* Tourism, Vol. 50 (2):165.

Johns, N., Pine, R. (2002): *Consumer behavior in the food service industry: a review,* Hospitality Management, HM: 407:1-16.

Kotler, P. (1996): *Marketing management,* Slovenska knjiga, Ljubljana:294.

Kuznesof S. et al. (1997): *Regional foods: a consumer perspective,* British Food Journal, 99, (6):199-206.

Muller, E. T (1991): *Using Personal Values to Define Segments in an International Tourism Market,* International Marketing Review, Vol. 8, (1):57-70.

Murphy P., Pritchard P. M., Smith B. (2000): *The destination product and its impact on traveler perceptions,* Tourism Management, Vol. 21:43-52.

Nickels, W. G., Wood, M. B. (1997): *Marketing - Relationship, Quality, Value,* Worth Publishers, New York:273.

Oliver, T. (2001): *The Consumption of Tour Routes in Cultural Landscapes,* Mazanec et al. (eds)., Consumer Psychology of Tourism, Hospitality and Leisure, Vol. 2., CABI Publishing, Oxon:273.

Pelham, A. M. (1997): *Market orientation and performance: the moderating effects of product and customer differentiatio,* Journal of Business & Industrial Marketing, Vol. 12 (5): 176-196.

Poon, A. (1998): *Tourism, Technology and Competitive Strategies,* CABI Publishing, Oxon:122.

Porter, E. M. (1996): *Il vantaggio competitivo,* Edizioni di Comunita', Milano:21.

Prentice R. C., Stephen F. W., Hamer C. (1998): *Tourism as Experience: The Case of Heritage Parks.* Annals of Tourism Research, 25 (1):1-24.

Priestley, G., Mundet, L. (1998): *The Post-stagnation Phase of the Resort Cycle,* Annales of Tourism Research, Vol. 25 (1):85-111.

Quijano-Caballero, C. (2002): *Cultural Tourism and Culinary Heritage.* Jelinčić, D. A. (ed.): Culture: A Driving Force for Urban Tourism - Application of Experiences to Countries in Transition. Institute for International Relations, Zagreb, 2002:64.

Richards, G. et al. (2001): *The Cultural Attraction Distribution System.* Richards Greg ed.: Cultural Attractions and European Tourism, CAB International, Oxon:74.

Rispoli, M., Tamma, M. (1995): *Risposte strategice alla complessita': Le forme di* offerta dei prodotti alberghieri, Giappichelli, G. (Ed.), Torino:26.

Sedmak, G., Mihalič, T., Rogelj, R. (2004): *What affects restaurant prices in tourist destinations? The case of the Slovenian coastal area,* Tourism 52 No 3:255-266.

Sedmak, G. (2004): *Tipična kulinarika področja kot element privlačnosti turistične destinacije,* International conference Encuentros - Strateška partnerstva za razvoj turističnih destinacij, Dnevi Turistice, Portorož.

Swarbrooke, J. (2002): *The Development and Management of Visitor Attractions,* Butterworth-Heinemann, Oxford.

Willmott, M., Graham, S. (2001): *The world of today and tomorrow: the European picture.* Lockwood, A., Medlik, S. (eds.): Tourism and Hospitality in the 21st Century, Butterworth - Heinemann, Oxford:31.

Maja Uran, Janja Jerman

Service Quality as a Competitive Advantage of the Slovenian Hotel Industry

This paper assesses the internal quality of service in the Slovenian hotel industry. It presents a management measurement tool that helps to assess the organizational service gaps that are preconditions for delivering service quality. If the service quality gaps are small, the hotel manager can use service quality as a source of competitive advantage. The presented theoretical model was constructed based upon the four organizational gaps of the Parasuraman's et al. service quality model, then redefined and reassessed. Data was gathered from a sample of 5000 questionnaires and analyzed with use of exploratory factor analysis and structural equation modeling. The results can provide useful guidelines for hotel management in terms of how to improve the service delivery process.

Key words: service quality model, internal service quality, organizational gaps, competitive advantage, multivariate analysis, hotel industry

1. Introduction

To create a sustainable advantage, firms seek to develop core competencies: unique combinations of processes, skills and/or assets (Dube, 2000;62-73; Roberts, 2001:116; Knowles, 1999:64). As competitors move more closely together in terms of product quality, it is the service quality, developed by these core competencies, which will be used more often to create a competitive distinctiveness (Zeithaml et al., 1990:149; Olsen et al., 1992:163; Harrington and Lenehan, 1998:15, Groenroos, 1990:123; Johns, 1999:205; Roberts, 2001:116). Service quality can be utilized in how a business produces and delivers its products and services; in how it manages its employees; and in how it builds a strong brand identity and reputation. It is a process that includes both the responsiveness of the service and the consistency of the service delivery. Firms that learn how to match service quality as an operational approach with their competitive methods can create a formidable and sustainable competitive advantage.

Mihalic and Konecnik (2000:537) have undertaken first major research into the competitiveness of the Slovenian hotel industry. As part of a larger research project, they made a selection of seventeen competitive advantage resources and asked Slovenian hotel managers (N=66; the response rate was 41 percent) to rank them from the most to the least important resource. The results are as follows:

1. Responsiveness and reliability of services

2. High quality of services

3. High local image and personal contacts

4. Service differentiation

5. Corporate image

6. High quality of post-sales services (after reservation)

7. Consulting before reservation

8. Sales channels

9. Brand recognition

10. Physical evidence and design

11. Internet presence

12. Quality of advertising, PR

13. Favorable terms of payment

14. Low prices

15. R&D expenditures

16. Market share

17. Central reservation system (CRS)

According to the results, non-price competitive advantage resources were highly ranked; the first three were directly linked to the concepts of service quality. Furthermore, they suggest that the only appropriate strategy for the average Slovenian hotel is Porter's strategy of production competitiveness based on product differentiation by using, primarily, service quality concepts.

The aim of this paper is to present a management measurement tool that helps to assess the organizational service gaps that are preconditions for delivering service quality. If the service quality gaps are small, the hotel manager can use service quality as a source of competitive advantage.

2. The Internal Service Quality Model – INSQPLUS

The service quality model was the framework for developing an internal service quality model - INSQPLUS. The original and extended Parasumaran's et al. model has 4 organizational gaps with 16 dimensions and 50 elements. Since the model wasn't tested in the tourism or hotel industry, or explored and confirmed using appropriate statistical methods, we decided to redefine and reassess the model.

The extended service quality model derived from Zeithaml et al. (1990:149) and based upon the in-depth review of over 300 literature units (Uran, 2003:220) on service quality management and an identification of the existing service quality models, we were able to illustrate the concept of internal service quality and attempted to develop an internal service quality model - INSQPLUS. The original model demonstrates how service quality emerges. The upper part of the model includes phenomena related to the customer, the lower part shows phenomena related to the service provider. According to the original model, five discrepancies between the various elements of the basic structure - quality gaps - can exist. These gaps are: gap 1 - the management perception gap (positioning gap), gap 2 - the quality specification gap (specification gap), gap 3 - the service delivery gap (delivery gap), gap 4 - the market communication gap (communication gap) and gap 5 - the perceived service quality gap (perception gap) (Zeithaml et al., 1990:150; Candido and Morris, 2000:216; McCarthy and Keefe, 1999:185; Powers, 2002:22, Lewis, 2003:109; Dunmore, 2002:39; Dale and Bunney, 1999:266; Johnson and Weinstein, 2003:205). Although the original model has 4 organizational gaps, the model INSQPLUS defines 5 internal (organizational) gaps that contribute to the final service quality gap between consumer expectations and service performance (perception gap). Because service quality in the original model is evaluated through the fifth or perception gap, we found it necessary to add a fifth gap to our internal service quality model – the evaluation gap. It is essential to have supervision and control or rather an assessment of the service delivery before the consumption of the services. The INSQPLUS theoretical model consists of 5 gaps - positioning, specification, service delivery, communication and evaluation - that have 26 dimensions with more than 100 elements. These quality gaps arise from inconsistencies in the quality management process. More specific, each gap is caused by certain factors. Every dimension of the original model was reassessed on the basis of literature.

Managers' perceptions of customers' expectations may differ from customers' actual needs and desires. This means that management perceives the quality expectations inaccurately. The positioning gap is due, among other things, (Zeithaml et al., 1990:149; Candido and Morris, 2000:216; McCarthy and Keefe, 1999:205) to:

- Inaccurate information from market research and demand analysis;

- Inaccurately interpreted information about expectations;

- Non-existent upward information from the firm's interface with its customers to management; and

- Too many organizational layers which stop or change the pieces of information that may flow upward from those involved in customer contacts.

Service quality specifications are not consistent with management perceptions of quality expectations. Thus, even if customer needs are known, they may not be translated into appropriate service specifications. So the specification gap arises as a result (Zeithaml et al., 1990:150; Groenroos, 1990:130; Lovelock, 1992:43; Candido and Morris, 2000:190; McCarthy and Keefe, 1999:205) of:

- Planning mistakes or insufficient planning procedures;

- Bad management of planning;

- Lack of clear goal setting in the organization; and

- Insufficient planning support for service quality from top management.

The service delivery gap is referred to as the service performance gap and means that quality specifications are not met by the performance in the service production and delivery process. This gap is due (Zeithaml et al., 1990:155; Candido and Morris, 2000:190; McCarthy and Keefe, 1999:205) to:

- Too complicated and/or rigid specifications;

- The employees do not agree with the specifications;

- The specifications are not in line with the existing corporate culture;

- Bad management of service operations;

- Lacking or insufficient internal marketing; and

- Technology and systems do not facilitate performance according to specifications.

The communication gap means that promises given by market communication activities are not consistent with the service delivered. The reason for the occurrence of this gap can be (Zeithaml et al., 1990:150; Candido and Morris, 2000:205; McCarthy and Keefe, 1999:205):

- Market communication planning is not integrated with service operations;

- Lacking or insufficient coordination between traditional marketing and operations;

- The organization fails to perform according to specifications, whereas market communication campaigns follow these specifications; and

- An inherent propensity to exaggerate and, thus, promise too much.

The perception gap (external gap) results when one or more of the previous five gaps occur. It means that the perceived or experienced service is not consistent with the expected service. For the company this ultimately results in bad word-of-mouth, negative impact on corporate or local image, decrease in revenues and lost business.

The gap analysis should guide management in identifying the reason(s) for the quality problem and discovering appropriate ways to close this gap. The internal service quality theoretical concept is presented in Figure 1. Through this concept internal service quality in the hotel industry can be defined as the process where a hotel company through its service positioning, service specification, service delivery, communication and service process evaluation according to customer expectations, is trying to achieve a level of hotel service, a service delivery process and a hotel company image that can satisfy customer expectations.

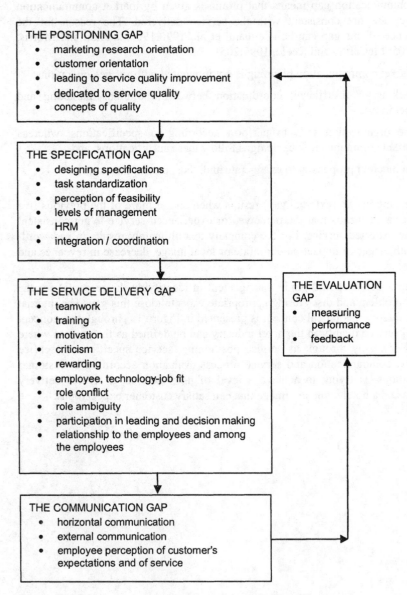

THE POSITIONING GAP
- marketing research orientation
- customer orientation
- leading to service quality improvement
- dedicated to service quality
- concepts of quality

THE SPECIFICATION GAP
- designing specifications
- task standardization
- perception of feasibility
- levels of management
- HRM
- integration / coordination

THE SERVICE DELIVERY GAP
- teamwork
- training
- motivation
- criticism
- rewarding
- employee, technology-job fit
- role conflict
- role ambiguity
- participation in leading and decision making
- relationship to the employees and among the employees

THE EVALUATION GAP
- measuring performance
- feedback

THE COMMUNICATION GAP
- horizontal communication
- external communication
- employee perception of customer's expectations and of service

Figure 1: The theoretical internal service quality model

To overcome the weaknesses of prior research, we conducted complex research to identify the representative structure and dimensions presented in the theoretical concept. Authors (Zeithaml et al., 1990:150) of the extended service quality model suggest that the model should be tested using appropriate multivariate statistical methods such as factor analysis. Measures of the theoretical construct affecting each gap can be viewed as indicators of that gap. Therefore, it is possible to recast the conceptual model in the form of a structural equations model. This model can be tested by collecting data on the indicators of the five gaps through questionnaires and analyzing the data using exploratory factor analysis, and then confirming the structure of the constructs using structural equation modeling.

3   Research Design and Data Analysis

The first research phase was dedicated to reviewing literature and defining a theoretical concept. However, research cannot be undertaken until research instruments, which usually have an American basis, have been developed and validated for local cultures and company sizes. For this reason we addressed the qualitative analysis of theoretical concepts by employing 15 experts from the Slovenian hotel industry. The results of this qualitative analysis formed the basis for the final operationalization of the measurement tools. This analysis also revealed that the theoretical concept cannot be tested directly, but needed be divided into two models. Firstly, an internal service quality model for hotel management with the following gaps: positioning, specification and evaluation. The 7-point Likert scale was used for this. Secondly, a model for the contact personnel comprised of the service delivery and communication gaps using a 5-point Likert scale.

Data were collected by mail survey conducted in Slovenia. Some questionnaires were handed over in person. Since there was no appropriate database of Slovenian hotel companies, one was created for this research. The questionnaire was addressed directly to all top executives and human resource managers in the entire Slovenian hotel industry. The human resource managers were asked to pass the questionnaires to all other managers in their hotels and to the contact personnel. Altogether 500 questionnaires were distributed to hotel managers and 4500 questionnaires to the contact personnel. Response rates in Slovenia may vary from 20-30 %. In case of this research, the response rates were within the limits:

- 33.6 % returned questionnaires from the hotel managers; and
- 23.1 % returned questionnaires from the contact personnel.

The gathered data were then analyzed using the chosen statistical methods. Exploratory factor analysis (EFA) supported by SPSS software was used to explore the gap structures. Structural equation modeling (SEM) supported by EQS software was used to confirm the gap structures. Each gap was explored individually using EFA until the appropriate structure was reached. This phase resulted in two integrated internal service quality constructs that were then tested using the SEM. Illustrating the individual results of EFA and SEM would exceed the aim and scope of this paper.

3.1   INSQPLUS for Hotel Management

The redefined internal service quality model for management, which was tested in the hotel industry, consists of the 4 dimensions (and 18 elements) shown in Table 1:

- marketing research orientation
- leading to service quality
- designing specifications
- performance measurement

Table 1: Internal service quality model for hotel management

DIMENSION	MEAN
Marketing research orientation	5,33
Leading to service quality	5,64
Designing specifications	3,90
Performance measurement	5,13
AVERAGE VALUE	5,00

The research revealed that the biggest problem of the Slovenian hotel industry in gaining sustainable competitive advantage based on non-price elements is a lack of the dedication of the hotel managers to defining hotel service specifications (mean 3, 90). On the other hand, this is the area where a lot can be improved to insure higher service quality. Still, the general assessment of the internal service quality of hotel management (average value 5,00) is that they understand the meaning of conducting market research, of service specification, of performance measurement and of implementing service quality models, but that, for various reasons, it is rare practice in the Slovenian hotel industry. The final conclusion is that Slovenian hotels can not use service quality as a source of competitive advantage due to their inappropriate approach to service positioning, service specification and service delivery process evaluation.

3.2   INSQPLUS for Contact Personnel

The redefined internal service quality model for contact personnel in the Slovenian hotel industry has the 4 dimensions (with 22 elements) shown in Table 2:

- rewarding and motivation
- participation in leading and decision-making
- relationship to the employees and among the employees
- employees' perception of service

Table 2: Internal service quality model for hotel contact personnel

DIMENSION	MEAN
Rewarding and motivation	2,75
Participation in leading and decision-making	2,70
Relationship to the employees and among the employees	3,38
Employees' perception of service	3,92
AVERAGE VALUE	3,19

If the internal service quality of hotel management could be described as mediocre, then - based on the average value from the internal service quality model for hotel contact personnel (mean 3, 19) - we can easily conclude that the latter is a disaster. Or that it is essential for Slovenian hotel managers to soon find an effective way of rewarding, motivating and including contact personnel in the decision-making process to insure successful business performance and higher service quality. As a result of this study, we can therefore definitely state that there will be no satisfied guests without satisfied employees in Slovenian hotels, and that there will be no means of gaining competitive advantage through a unique service production process, which is ultimately what service quality stands for. The final conclusion is that Slovenian hotels can not use service quality as a source of competitive advantage due to inappropriate approach to service delivery and communication.

4. Conclusion

We believe that the INSQPLUS model and its instruments could be a useful tool for measuring internal factors - internal service quality. Researching the internal gaps can help hotel managers to identify areas of improvement and to find an effective way of managing hotel human resources. Assurance of the high standard of service delivery will lead to higher service quality, which is commonly seen as being the best source of (sustainable) competitive advantage.

References

Candido, C. J. F and Morris, D. S (2000): *Charting service quality gaps*. Total Quality Management. Jul, vol.11, issue 4-6:216.

Dale, B. and Bunney, H. (1999): *Total Quality Management Blueprint*. Malden: Blackwell Publishing:266

Dunmore, M. (2002): *Inside – Out Marketing*. London: Kogan Page Ltd.:39.

Gronroos, C. (1990): *Service management and marketing. Managing the moments of truth in service competition*. Lexington: Free press/ Lexington Books:123-130

Harrington, D. and Lenehan, T. (1998). *Managing quality in tourism*. Dublin: Oak Tree Press:15.

Johnson, W. C. and Weinstein, A. (2003): *Superior Customer Value in the New Economy*. Boca Raton: CRC Press:205.

Johns, N. (1999): *Quality management*. In Brotherton, B.(Ed.)(1999) The handbook of contemporary hospitality management research. Chichester: John Wiley&sons, Ltd.:333-350.

Knowles, T. (1999): *Corporate strategy for hospitality*. Harlow, Longman:64.

Lewis, B.R. (2003): *Managing service quality*. In Dale, B. G. (Ed.): Managing Quality. Malden, Blackwell Publishing:109.

Lovelock, C. H. (1992): *A basic toolkit for service managers*. In Lovelock, C. H. Managing services: marketing, operations and human resources. New Jersey, Prentice-Hall:43.

McCarthy, P. M. and Keefe, T. J. (1999): *A measure of staff perceptions of quality*. Journal of Quality Management, 4/2:185-207

Mihalič, T. and Konečnik, M. (2000): *"Slovenian hotel industry competitiveness before and after EU entry"*, In the Tourism and hospitality management: trends and challenges for the future-International congress HOTEL 2000, Proceedings, Faculty of Tourism and Hospitality Management Opatija, Croatia: 537-557.

Olsen, M. D, Ching-Yick Tse, E. in West, J. J.(1992): Strategic management in hospitality industry. New York, Van Nostrand Reinhold:163.

Powers, T. (2002): Marketing hospitality. Toronto: John Wiley & Sons:22.

Roberts, G. N. (2001): *Competitive Advantage of Service Quality in Hospitality, Tourism and Leisure Services.* Kandampully, J. et al. (eds.): Service quality management in Hospitality, Tourism and Leisure. Binghamton, The Haworth Press:116-118.

Uran, M. (2003): *The service quality as the differentiation strategy for gaining competitive advantage of the Slovenian hotel industry*, Ljubljana: Faculty of Economics, University of Ljubljana, doctoral dissertation:220.

Zeithaml, V. A. et al. (1990): *Delivering quality service*, New York: The Free Press:149.

Tadeja Jere Lazanski

System Dynamics Models to Support Decision-Making in Tourism

Abstract

Tourism is a type of inter-organizational system with global and local properties. It is a complex system with certain structural and behavioral properties. Problems are defined softly and phenomena are uncertain. There are demands for fast and integrated decisions to satisfy guests as service users and tourism organizations as service providers. There are many different methodologies and methods, which master softly structured problems. Here one encounters methods of system dynamics and system thinking. These tools were first brought into force in education and training in the form of different computer simulations and later as tools for decision-making and organizational re-engineering. System dynamics models are essentially simple and can serve only to describe the activity of basic mutual model values determination and the so-called mental model, which is the basis for causal connections among model variables. Thus, a connection between qualitative and quantitative models will be restored.

Keywords: systems, system dynamics, tourism, modeling, causal-loop diagrams, decision-making

1. Introduction

The success or failure of a particular policy initiative or strategic plan is largely dependent on whether the decision-maker truly understands the interaction and complexity of the system he or she is trying to influence. Considering the size and complexity of systems that public and private sector decision-makers must manage, it is not surprising that the "intuitive" or "common sense" approach to policy design often falls short of, or is counterproductive to, desired outcomes. Tourism is a relatively recent socio-economic activity, encompassing a large variety of economic sectors, players and academic disciplines. Researchers, who have estimated the economic effects of tourism on regional and national economies, modeled tourism as an exogenous activity (Alavapati and Adamovicz, 2000:188). The complexity of its composition makes it inherently difficult to develop universally acceptable definitions, which would help to describe it conceptually.

The rapid development of information technology, especially the Internet, has also caused changes in organizational systems. Work methods and boundaries of, and the organizations within the system have changed. The main properties and driving forces of contemporary organization are information processing, decision-making and learning. Typical of such organizations, the market environment for tourism is becoming increasingly like this.

Turbulence in the world market is demanding flexibility and fast response times from the entire service industry. It requires decisions that frequently reflect opposing interests. Thus, an excellent methodological approach to these problems is urgently needed. We encounter the methods and tools of system dynamics and system thinking, which became common management tools in the 1990s. Since system dynamic models are essentially simple, one must find a compromise between simplicity, limited usefulness and complexity. At present, the most intensive research efforts are focusing on a combination of simulation methods and expert systems (Kljajić, 2000:293; Dijk, 1996:106). One will pay attention to the methodology for parameter model values determination and so called mental model, which is the basis for causal connections among model variables. Qualitative research provides a crucial perspective that helps scholars to understand phenomena in a wider sense than merely from a positivist perspective (Riley and Love, 2000:164).

This work attempts to critically analyze and evaluate the methodologies and apply them to a tourism development model.

2.   Systems, Systems Dynamics and Tourism

In general we can say that a system is comprised of interrelated components, connected together in order to facilitate information, matter and energy flows. The central concept system embodies the idea of a set of elements connected together, which form a whole rather than properties of its components parts (Checkland, 2000:3).

In operations research and organizational development, organizations are viewed as human systems comprised of interacting components such as sub-systems, processes and organizational structures. Many researchers have joined in the ongoing discussion between system dynamicists as to whether simulation experiments without modeling can yield substantial gains for the user.

System dynamics appear to be an appropriate and effective tool for creating the underlying formal simulation model (Größler, 2001:72). They are a powerful methodology and computer simulation modeling technique for framing, understanding, and discussing complex issues and problems. A system dynamics approach can improve the intuitive understanding and illuminate the relationships among the various parts of the whole system (Brokaw and Jambekar, 1998:205).

System dynamics address the behavior of a system over time. A critical step in examining a system or issue is to identify its key *patterns of behavior* - what we often refer to as "time paths". System dynamics provide the basic building blocks necessary to construct models that teach us how and why complex real-world systems behave the way they do over time. A tourism system is a system concerning the tourism domain (travel, tourists, resorts, hotels, restaurants, etc.) that can be described by various components where different processes take place (economical, psychological, sociological, physical, etc.). Thus we obtain a link between a system and its dynamics. System dynamics therefore can be introduced to combine both 'hard' quantitative dimensions and the 'soft' qualitative dimensions. Forrester (1973) defined system dynamics as the investigation of the information feedback characteristics of systems and the use of models for designing improved organizational forms and management policies.

System dynamics can be applied to any dynamic system, over any time and spatial scale (Sterman, 2000:42). It is a method for enhancing learning in complex systems and for developing management flight simulators, often computer simulation models, to help us learn about dynamic complexity, understand the sources of policy resistance, and design more effective policies.

Combining elements of these viewpoints, system dynamics may be defined to include:

a) Methodology for understanding complex problems where there is dynamic behavior (quantities changing over time) and where feedback has a significant impact on system behavior.

b) Framework and rules for the qualitative description, exploration and analysis of systems in terms of their processes, information, boundaries and strategies, facilitating quantitative simulation modeling and analysis for the design of system structure and control.

c) Rigorous study of organizational problems from a holistic or system perspective using the principles of feedback, dynamics and simulation.

3.   Qualitative Modeling of Complex Systems

The structural properties of complex systems are: local interactions, non-linearity, feedback and openness (difficult to find boundaries). Its behavioral properties include: emergency, self organization and adaptiveness. This means that tourism is a complex system with its huge quantity of data to manipulate, low quality of data (uncertainty, measurement, errors, missing data), different spatial and temporal scales (from seconds to years, from local to global), dynamic and stochastic behavior, and its numerous interfaces to many disciplines/domains.

Model building in tourism seeks to understand a complex relationship and to aid the management of a place or process. For example, econometric forecasting of tourism flows aims to help estimate future numbers of tourists in order to permit informed decisions.

Modeling requires a disciplined approach and an understanding of business, skills developed through study, and experience (Sterman 2000:81). It is an attempt to identify key variables in a situation and the relationship that exists among them.

The model is kept as simple and general as possible to emphasize the interactions between the economy and the environment, and their impact on tourism (Alavalapati and Adamovicz, 2000:188). The choice of model depends on the system we want to examine and the aims of our research.

Every development strategy, including tourism strategies, needs to be presented as a whole, with all its strengths, weaknesses, future expectations, and using different methods. The models are simplifications, abstractions of features deemed to be important, and there can be no guarantee that they will be valid. But, used sensibly, models and modeling approaches provide one way of managing risk and uncertainty (Pidd, 1996:29).

At manifestation level, the system is described as it appears, instead of as it is. By definition, we anticipate that the system consists of elements and is greater than its parts. An element is the smallest part of the whole necessary for system description, which can't or won't be divided further. The essence of the elements is very important from the epistemological point of view. From the general point of view a system is defined by the formula:

$$S = (E, R), \quad (1)$$

Where $e_i \in E \subset U, = 1,2,..n$, represents the set of elements (2)

$R \subseteq E \times E$ The relation between elements, and U common domain

Modeling involves describing our experiences by using one of the existing languages in the framework of a certain theory. In this way, our experiences also become accessible to others: they may be proven, confirmed, rejected, broadened or generalized. This paradigm can be stated (Kljajić, 1998:864) using a triplet (O, S, M). O represents the real object, original, independent from the observer, while S represents the researcher (subject) or an observer with his knowledge, and M the model of the object (Kljajić, Jere Lazanski, 2001:216). It is very similar to Peirce's philosophy of three universal, phenomenological categories, Firstness, Secondness and Thirdness, whose "prelogical" characters always appear when we engage in "honest, single-minded observation of the appearances" (Brent, 1998:352).

A description of a system depends on exactly defined goals and the researcher's point of view. From a methodological point of view, which is more important, we can only understand complex systems as a whole, i.e. its complete portrait within its surroundings. An external world exists independently from the observer and isn't directly observable, so we set up simplified models to represent it. The relation between the observer S and the object O is crucial for the cognitive method. The observer is a man, with all his cognitive qualities, while the research object is the manifested world, which exists by itself, regardless of how we describe it. In this case, the object and the system have the same meaning. The third article of the triplet M is the consecutive one and represents a model or a picture of the analyzed system O. The $O \leftrightarrow S$ relation in Fig. 1 indicates the reflection of human experiences to concrete reality. This cognitive consciousness represents our mental model. The relationship $M \leftrightarrow S$ represents the problem of present knowledge, respectively the translation of the mental model into the actual model. The $O \leftrightarrow M$ relation represents the phase of model validation or proof of alignment between theory and practice, which make it possible to generalize experiences into rules and laws. The $S \rightarrow O \rightarrow M$ relationship is nothing else but an active relation of the subject in the phase of the object's cognition.

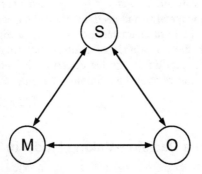

Figure 1: Subject in a modeling process (Source: Kljajić, 1998:869)

The $M \rightarrow O \rightarrow S$ relation is nothing more than the process of learning and generalization. Just as we can talk about complexity of object O, we can also talk about state, goals and estimations S, about homomorphous and isomorphous connection between a model and the original.

The relativity of knowledge, theory and scientific models is obvious. These develop constantly through time in elegant interaction with civilization processes. The characteristics of an observer can be included as a sub-domain of a product:

$$O \subseteq I \times C \times Z \times T \quad \text{(3)}$$

Where

I.....represents the starting-point of a research observation

C...goal of the observer

Z...knowledge of the observer

T... technology or tools, which the observer uses

The above mentioned is quite clear by itself. But the need to be aware of it when describing a complex system is very important and not obvious.

The work presents the connection between the principle process of modeling and the communication model among participants of complex systems (Kljajić, Jere Lazanski, 2001:218), and shows that every model context depends on Peirce's triad philosophy, where all matters divide on the base of a trichotomy and triad relationship into three categories: *firstness, secondness and thirdness*. In the case of a trichotomy of chance, reality and need, he divided chance as a first, reality as a second and need as the third characteristic of a matter with regard to priority. Similarly, in terms of the importance of matters, he ranked quality first, fact second and habit (law, rule) third. The list continues infinitely, and we can use Peirce's method to divide relationships among elements according to the triad principle.

4. Concept of a Decision-Making Model in Tourism

Decision-making processes are the focal point of any system development, including tourism development. The decisions that include a wide variety of financial, technical and logistical resources require simulation by decision-makers before they are put into action or production (Kljajić et al., 2002:202). The decision-makers in the decision process are supported by a simulator, which enables decision testing to be an integral part of a business plan. As in all organizational systems, subjective factors, such as human skills and creativity, play an important role in successfully solving problems. Teamwork plays the most important role in the process of achieving optimal decisions.

Complex systems are usually presented as a black box, which has certain inputs (people, information and means) and outputs (services, final products). Thus it is delimited to the environment during the problem defining phase. Organizational systems have inner structures, which transform variables (input quantities) into output functions (output quantities)

There is another characteristic of an organizational system, which is defined as the causal loop concept and is present in all systems, which connect input and output elements in a way that output elements can influence back to input. This is extremely important, since this is how regulation processes function.

Organizational systems, which include tourism systems, are dynamic. Regulation is necessary but far removed from being sufficient. The strategic vision of a development and system environment to influence prediction is most important. For this reason organizational systems can be defined slightly differently, to emphasize the inner causes of system behavior. Usually these are called management subsystems. A general model of a goal-oriented system is defined using a pair (P, D) and presented in Fig. 2. P represents the managing process in tourism, D represents the managing subsystem. Loop $P \rightarrow Y \rightarrow D \rightarrow U \rightarrow P$ represents feedback information, which functions on the cause-effect principle; therefore we can call it reactive control. Such control is sufficient for small incidents. Information from the environment is needed to make decisions in organizational systems. The chain $X \rightarrow D \rightarrow U \rightarrow P$ provides feed forward information, which represents the anticipation of the future state of the environment. It is an important part of the strategy of goal-oriented systems.

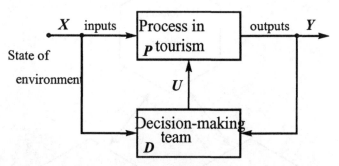

Figure 2: General model of a tourism-oriented system (Source: Jere Lazanski, 2002:54)

Man is part of a managing system and the one who creates goals and bears responsibility for the system development. His knowledge and awareness influence inter-relationships, the organization of technical and natural parts of subsystems for achieving quality goals, and the functioning of the system.

There is no absolute solution, since systems are open, new goals are defined, and the reality of an environment needs to be appreciated. Past experience and future anticipation enable the development and growth of these systems. Simulation is one of the best options for developing visions in tourism control.

A simulation model included in a user-friendly simulator enables the decision-makers (policy makers, system analysts and stakeholders) to analyze different simulation scenarios. The implementation of the Group Support Decision System (GDSS) enables the participants in a decision-making process to test different business scenarios and share a common view when considering a problem. Later the indirect effects of testing scenarios on the environment can be understood without risking their direct implementation into a production process. Use of simulation as a base for certain decision gives a new value to anticipated information, which facilitates the adapted nature of the decision-making process. The main paradigm of problem solving via simulation is shown in Fig. 3. It represents the relations among the participants (decision-makers), the business system and its simulation model. The simulation approach seems to be one of the better methodologies used to achieve and anticipate information for decision-making in enterprise system.

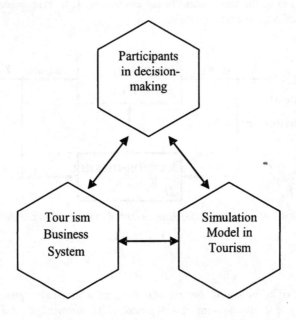

Figure 3: Simulation methodology determination with participants' involvement (Source: Jere Lazanski, Kljajić, Škraba 2002:121)

Roughly speaking this comprises the concepts of state, goal, alternative criteria and a state of nature, which are connected in a dynamic model that interacts with the decision-making groups. The business process was designed on the concept of the state-variable approach. It signifies a quantity, which represents the main entity relevant for decision-making at top and operative levels. The system for decision assessment has been organized in two hierarchical levels. The model at the top is used for assessing enterprise strategy. At the bottom level the model is used for discrete event simulation, which is necessary for operation planning and testing the service performance. The concept of state is convenient for achieving harmony among the different levels throughout the whole system. In a practical sense, this means that when the discrete-event process is considered, variables are considered as entities at the level and rate in the system dynamics when the process is considered continuous. A conceptual view of the proposed approach is shown in Fig.4.

The core of a business system simulator is a simulation model. Solving problems using simulation models follows standard steps: state analysis, development of a causal loop diagram, writing of the model's equations and model implementation. Particular scenarios that form and determine the tourism market in a certain environment are tested on a simulation system.

A simulator is connected to the GSS (Group Support System). The participants using GSS work directly with the system simulator. A system simulator is connected directly to a database necessary for simulation model activation. Simulation results are evaluated both with the group decision-making support system and with the expert system. The understanding of the system increases throughout this entire process. With the described model, the experimental loop on a simulation model has been finished with the help of a system simulator and scenario ranking.

The implementation of the above-described methodology is shown in Fig. 4. It presents the principle scheme of a simulation system for decision assessment in tourism. Modeling and scenario determination represent a knowledge-capturing process in the form of the structure and behavior of the model. Once the model is defined and validated, experimentation with different scenarios is possible. The expert group determines the set of different scenarios, which represents the possible future action in the real system. The results gathered as the output of the model are evaluated using the multi-criteria evaluation function.

Information feedback provides the expert group with the possibility to creatively determine a new set of scenarios and multi-criteria evaluation functions relating to the given situation. Simulated and actual performances of the system are compared in order to adapt the strategy according to changes in the environment.

Implementation of the simulation system enhances the learning process (David and Richardson, 1997:111). Results are continuously communicated to the expert group, providing an informational feedback loop in the learning process, which has a significant impact on the decision process, as preliminary analysis has indicated.

Figure 4 shows the interaction between the business system and the people involved in it - the participants in a decision-making process and simulation model.

The participants in a decision-making process are a part of the tourism business process. The roles of decision makers are: the technical authority, which orders a project (e.g. tourism authority), the political authority which approves/rejects a project (e.g. a local government), the system analyst who develops the project and the stakeholders (local interest groups).

The model can be used as a basis for accepting business decisions.

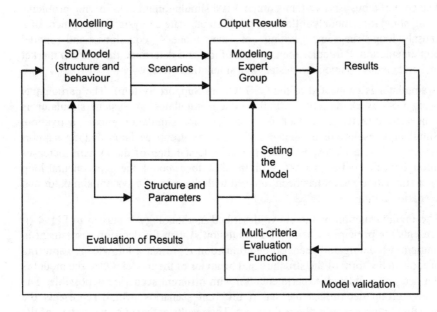

Figure 4: The principle scheme of simulation methodology for decision-making support (Source: Jere Lazanski, 2005:61)

During the experiment, a group decision-making tool is used: the Ventana Corporation's A Group Systems. Work with this tool is anonymous, which enables a greater flow of ideas and reduces unwanted influences. The participants become more relaxed since no one knows where the ideas are coming from, which motivates creativity; this simply would not be the case with more classical work approaches. The work time decreases and the efficiency of participants increases. The final result is better as the decision becomes a group decision with which conflict between polarized groups is minimized and a consensus is achieved for the development of further actions. Present opportunities and future needs for this kind of decision-making system must be mentioned. During vacation, a holidaymaker can play a dual role and might have to work creatively using remote systems. For this purpose a tourist organization must provide him with high quality information systems for the Internet, Intranet, teleconferences, group decision support systems, GSM and other means. An important role can be given to expert systems and virtual wizards. A decision-making support system must satisfy both tourism service users and tourism service providers:

1. Tourism industry decision-making for the provision of rational and excellent service together with the participation of holidaymakers and
2. Enabling a system for global decision-making and different working areas (for holidaymakers)

Therefore, we can include tourism organizations into inter-organizational systems with local and global elements.

5. Developing System Dynamic Models in Tourism

In tourism management it is essential to develop a system approach to addressing the impact of economic and social factors interacting in a region. A core part of designing a tourism system is recognizing the basic causes of problems and evaluating the consequences caused by management answers. There are plenty of views, which research different areas in tourism. System dynamic models allow computing scenarios to assess the possible implications of strategic situations. These are not merely hypothesized plausible futures, but computed by simulating changes in strategy and the business environment. (Georgantzas, 2003:179) System researchers will discover that tourism is a "black box", which means tourism is a system with inputs and outputs, and interactions among single subsystems. The basic qualitative strategic goals of national tourism (in the case of Slovenia) are shown in Figure 5:

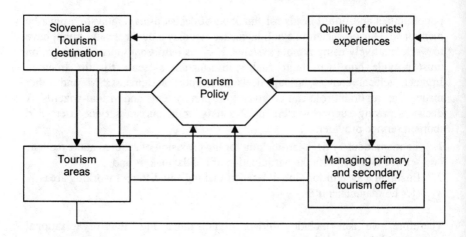

Figure 5: Slovene tourism system diagram (Source: Jere Lazanski, 2002:30)

Fig. 6 presents a tourism simulation model to explain the methodology described above. Many equivalent illustrations of systems, which are appropriate to computer simulation, exist. In business management simulation a methodology of system dynamics has emerged, which was suggested by J. Forrester (Forrester, 1973). All discussions relating to modeling lead to the same conclusions. There are few minor differences among the graphic illustrations of elements and their interactions. Forrester suggests that the SD method has some semantic advantages for those users who have less experience with formal methods. The methods are equivalent; the only difference lies in the fact that the influential diagram more closely resembles graphs respectively qualitative models. To illustrate the equivalency of modeling methods, one must present a CLD (causal loop diagram) of dependency among environmental attractiveness, number of tourists and infrastructure investments.

Following a definition of system equations (1) the basic elements of a system can be defined:

E_i = environment, attractiveness of the environment, number of tourists, investments, infrastructure, crowding, and their interconnections

R = environment, attractiveness of the environment, number of tourists, which creates a simplified model

Understanding a process is a base for R set connection. It can be described as follows: a preserved environment (+) positively impacts its attractiveness as a tourism area (+), which influences the number of tourists, (+), the number of tourists influences the growth of investments in infrastructure and a culture of quality of life (+). On the other hand: more tourists (+) cause environmental damage (-), which reduces the attractiveness of a tourism area. At the same time, crowding (+) causes detours, traffic jams, drivers' nervousness, accidents, anger and regrets about deciding to spend holidays in this kind of area. (-) These qualitative descriptions illustrate what must be taken into account. If we connect a set of elements Ei based on their descriptions with a pointed arrow in the same direction and mark this with a symbol (+), and the opposite with a symbol (-), we get an influential diagram respectively qualitative model of our simplified system, as shown in Fig.6. The model illustrates that there is one basic circle (-) of causal loop, the growth in the number of tourists and restrictions to growth caused by infrastructure and damage to the environment. In a vision of tourism strategy development, the development must be predicted as a whole in order to avoid limitations. If only one element in the reinforcement circle, consisting of investments and environmental preservation, starts to decline (-), it causes all the other elements to decline (the number of tourists decreases).

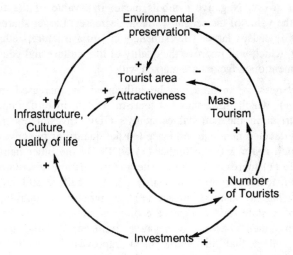

Figure 6: CLD (causal loop diagram) dependency of environmental attractiveness, number of tourists and infrastructure investments (Source: Jere Lazanski, 2002:64)

System dynamics are particularly useful in understanding the links between the qualitative and the quantitative aspects of tourism management. System dynamics modeling employs a set of techniques that allow both a quantitative and realistic representation of variables that are typically perceived to be qualitative.

6. Tourism System as a Part of National Strategy

Tourism is a part of national development policy and it must be discussed as a whole – together with economics, infrastructure and education. System dynamic methodology incorporates all the development perspectives.

The causal loop diagram in Fig. 7 presents a macro tourism model as a part of the national economy, where a share of tourism is visible in terms of GDP and the realized goals of a development strategy. Causal loop circles and interactions of single branches are visible. Places described with variables represent elements of system state, arrows among them show a trend of influence of single elements. A symbol at input or output of an element shows a trend of change. Positive symbols mean: if a value of the first variable grows, then the value of the second variable also grows. Negative symbols mean: if a value of the first variable grows, then the value of the second variable decreases. Larger shares of tourism in GDP will cause larger investments in infrastructure, education and development, which will improve the quality of life, culture and education level. All of these mentioned factors are marked with (+).

The attractiveness of an area can be influenced by increased infrastructure investment (+), which influences (+) tourism, which, in turn, impacts (+) the share of tourism in the national economy's GDP. Investments in education, research and development can influence (+) the quality of the environment and industry, which again leads to higher (+) GDP. On the other hand, improved quality of life (+) increases the birth rate and population of a state, (+) which negatively impacts environmental quality (-). The culture and literacy of the surroundings has a positive impact (+) on agricultural quality, while the agriculture of a state has a negative ecological impact (-) on environmental quality. Employment has (-) a negative influence on the quality of an environment, albeit that this employment improves the quality of life (+). Industry also has a negative (-) impact on the environment.

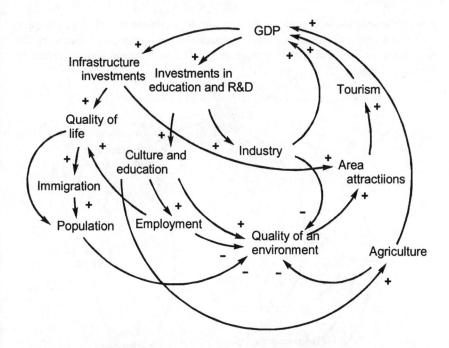

Figure 7: Causal loop diagram (CLD) of a tourism system as a part of national strategy (Source: Jere Lazanski, 2002:65)

The diagram illustrates a national tourism market, which has a positive influence on the tourism infrastructure respectively secondary tourism offer (+), good infrastructure has a positive effect (+) on the tourism product, the tourism product positively affects the attractions offered by national tourism (+). The national tourism market has a positive influence (+) on tourism programs, which positively affects (+) the animation and leisure time of tourists; this has a positive effect (+) on the attractions offered by Slovenian tourism. Attractions cause growth in the number of tourists (+) in a local tourism market, crowding detracts from the quality of an attraction (-). Secondary tourism offers have a negative impact (-) on environmental preservation, which lowers the attractiveness of an area (+). Circles of positive feedback mean development, but it must be emphasized that every decrease in a circle is followed by a drop in growth. If the environment is ecologically damaged, this impacts the attractions (-), the number of visitors decreases (-) and the anticipated financial return from this aspect of tourism decreases. Investments in tourism also lose popularity (-).

For this reason we will improve the tourism development model from Figure 6 by incorporating the other aspects of sustainable development in terms of the national economy and global surroundings. The new causal loop diagram is shown in Figure 8:

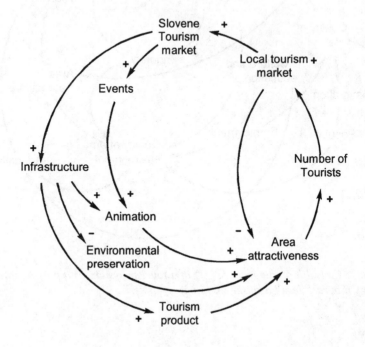

Figure 8: CLD of integral tourism development policy from a national tourism market perspective (Source: Jere Lazanski, 2002:66)

This diagram shows a tourism market, which has a positive influence on the tourism infrastructure and secondary tourist offer (+), the good infrastructure has a positive impact on the tourist product (+), the tourist product influences the attractions offered by tourism (+). The national tourism market has a positive influence (+) on tourism programs, which positively influence animation, animation then has a positive influence as well (+) on the attractions offered by national tourism. Attractions cause growth in the number of tourists (+) at local level (+), the number of tourists causes the quality of attractions to decrease (-).

The tourism infrastructure has a negative influence on environmental preservation, (-), which detracts from the attractiveness of the tourism area. (-). Positive causal loop circles mean development, yet it must be said that every negative impact is followed by a decline in growth. For example, if the environment is ecologically poor, this negatively influences area attractiveness, producing a chain reaction of decreasing numbers of tourists, reductions in tourism investments and financial losses.

7.   Simulation Model

The causal loop diagrams we have described represent qualitative models of Slovenian tourism. They are followed by system dynamic models, which are actually simulation models. The difference between a causal loop diagram and a system dynamic model lies in the quantity of parameters and concrete data needed for simulation, which are gathered in SD. A project aimed at creating a development strategy grows with the discussion of different scenarios. Figure 9 presents an example of a simple stock and flow structure. In system dynamics modeling, dynamic behavior is thought to arise as a result of accumulation. More precisely, this principle states that all dynamic behavior in the world occurs when flows accumulate in stocks. According to Sterman (Sterman 2000:194) the flows are functions of stock and other state variables and parameters.

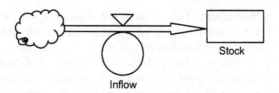

Figure 9a - Example of a simple stock and flow structure

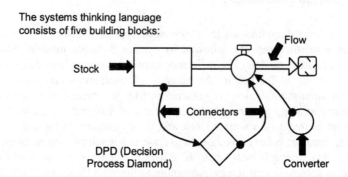

Figure 9b - Example of a simple stock and flow structure

Figure 9a and 9b (Copyright: Ithink-Issee systems, 2000): Examples of a simple stock and flow structure

In metaphorical terms, a stock can be thought of as a bathtub and a flow as a faucet and pipe structure that fills or drains the stock as shown in Figure 9. The stock-flow structure is the simplest dynamic system in the world. According to the principle of accumulation, dynamic behavior arises when something flows through the pipe and faucet structure and collects or accumulates in the stock. In system dynamics modeling, both informational and non-informational entities can move through flows and accumulate in stocks. Stocks usually represent *nouns* and flows usually represent *verbs*, they do not disappear if time is (hypothetically) stopped (i.e., if a snapshot were taken of the system); Flows do disappear if time is (hypothetically) stopped and they send out signals (information about the state of the system) to the rest of the system.

Figure 10 shows a system dynamics model depicting the interactive dependence among environment attractiveness, number of tourists and investments in infrastructure. In the experiment, this model is defined as the "real world system." This is followed by the creation of an exact copy of the "real world system". The "model" is perfect in the sense that its nonlinear stock-flow-feedback structure, its parameters, its distribution of random variates, and its initial values are identical to those of the "real world system." The "model" is thus more perfectly specified than any actual social system model could ever be in the true real world. Stocks or Levels show a variable type and a model object in Powersim models, used to represent the state variables of a system. Levels accumulate connected flows. Array Stock has one dimension with different elements, and flows in a Powersim model represent the transport of quantities to, from, and between levels, whereas connectors are links to establish an influence from one variable to another

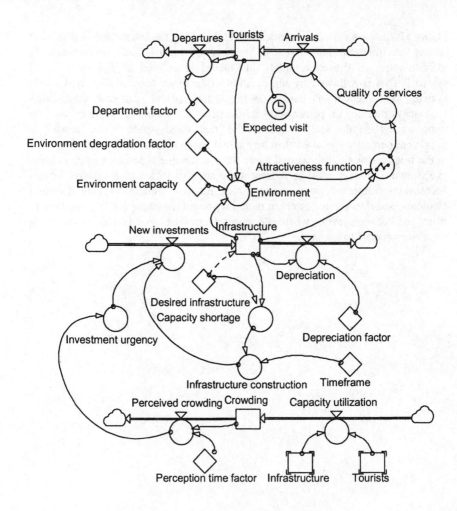

Figure 10: SD diagram of a tourism simulation model for decision-making support (Source: Kljajic, Legna, Skraba 2002:205)

Implementation of the system enhances learning processes (David and Richardson, 1997:120) as is also proven in real cases. Companies can use simulations to test tactical decisions and experiment with marketing or product development strategies. From a logical point of view, the testing of a theory depends upon basic statements whose acceptance or rejection, in turn, depends on our decisions. (Popper, 2002:91) The purpose of simulations is to help people understand the basics of business and, in particular, the financial implications of various decisions.

7.　Conclusion

This work has attempted to present a concept for the meaningful use of a system of simulation methods, techniques and expert systems as a functional part of decision-making and participation in tourism. It has discussed the use of an excellent methodology for tourism as a complex system – system dynamics and modeling within this framework.

Our work so far reveals the following advantages:

* System dynamics modeling over more traditional statistical correlation modeling;
* The graphical interface visualizes the relationships between key variables;
* Qualitative data, which is important in the decision-making process, can be readily incorporated into the model;
* The model readily integrates database-type pavement management systems, significantly enhancing their potential for scenario analysis;
* The causal-loop diagrams, which were developed in order to optimize the decision making process
* Users gain confidence in identifying "good plausible settings" of the model parameters, from which subsequent "what-if" analysis can proceed.

Further work is required on the incorporation of qualitative 'social and political' feedback mechanisms and also on the incorporation of work practice parameters into the prioritization function.

References

Alavalapati R. J., Adamovicz W. L., (2000): Tourism Impact Modeling For Resource Extraction Regions, Annals of Tourism Research, Vol.27, No. 1:188-202, Pergamon-Elsevier Science.

Brent, J. (1998): *Charles Sanders Peirce: A Life.* Indiana University Press, Indiana:352.

Brokaw, A. J., Jambekar, A. B. (1998): *"Using System Dynamics Modeling to Investigate Public Policy Issues for Snowmobile Tourism."* Proceedings Atlantic Marketing Association, Fourteenth Annual Conference, Savannah, Georgia:203-209.

Checkland, P. (2000): Systems Thinking: Systems Practice, Wiley, Chichester:3.

David, F. A. and Richardson, G. P. (1997): Scripts for group model building, System Dynamics Review, Vol. 13, No.2:107-129.

Dijk J. N. et al. (1996): Visual Interactive Modeling with Sim View for Organizational Improvement. Simulation, 67. No. 2:106-120.

Forrester, J. W. (1973): Industrial Dynamics, The M.I.T. Press:13.

Georgantzas, N. C. (2003): Tourism Dynamics: Cyprus' Hotel Value Chain and Profitability. System Dynamics Review Vol. 19, No. 3:175-213.

Größler, A. (2001): "Musings about the Effectiveness and Evaluation of Business Simulators, in J. H. Hines et al. (eds.): Proceedings of the 16th International Conference of the System Dynamics Society:72.

Jere Lazanski, T. (2002): *Qualitative Modeling of Complex System in a Frame of System Dynamics*: Masters Thesis, University of Maribor, Faculty of Organizational Sciences, Kranj:30-55.

Jere Lazanski, T. (2005): *System Approach to a Context Dependent Modeling of Complex System and a Problem of Validation*: Doctoral Thesis, University of Maribor, Faculty of Organizational Sciences, Kranj:55-66.

Jere Lazanski, T. et al. (2002): A concept of a multi-criteria decision-making system in tourism, using models of system dynamics. *Tour. Hosp. Manag.*, vol. 8, no. 1/2:117-125.

Kljajić, M. (1998): Modeling and Understanding the Complex System within Cybernetics. Ramaekers, M.J. (ed.), 15th International Congress on Cybernetics, Association International de Cybernetique, Namur:864-869.

Kljajić, M. (2000): Simulation Approach to Decision Support in Complex Systems. International Journal of Computing Anticipatory Systems, Vol.5, Dubois, D. M. (ed.), CHAOS, Liege, Belgium:293-304.

Kljajić, M., Lazanski, T. J. (2001): System Approach to Modeling of complex System: (with special regards to inter-organizational systems). In: DUBOIS, Daniel M. (Ed.). CASYS '01: Liege: CHAOS cop.:216-218.

Kljajić, M. et al. (2002): System dynamics model development of the Canary Islands for supporting strategic public decisions. In *Proceedings of the 20th International Conference of the System Dynamics Society, Palermo, Italy, July 28 - August 1, 2002*. [Albany]: System Dynamics Society:202.

Pidd, M. (1996): *Tools for Thinking: Modeling in Management Science*. John Wiley & Sons, Chichester:29.

Popper, C. (2002): *The Logic of Scientific Discovery*, Routledge, London:91.

Richardson G. P. and Andersen D. F. (1995): Teamwork in group model building. System Dynamics Review, Vol. 11, No. 2:113-137.

Riley, W., Love, L. (2000): The State of Qualitative Tourism research, Annals of Tourism Research Vol. 27, No. 1:164-187, Pergamon-Elsevier Science.

Sterman, J. D. (2000): *Business Dynamics: Systems Thinking and Modeling for a Complex World*. Irwin McGraw-Hill, Boston:42-81,194.

Internet source: http://www.hps-inc.com (isee systems, Inc. 46 Centerra Pkwy, Suite 200, Lebanon, NH 03766-1487)

Boris Bukovec, Rok Ovsenik

Fundamentals of the New Paradigm of Quality Organizational Change Management

Abstract

In Europe, we are currently witnessing strong globalization processes, which are giving rise to a host of multi-layered problems and opportunities. With its focus on increased competition, the European Union touches upon a broad spectrum of factors connected also with change management, quality and creativity. This turbulent environment offers new opportunities that can be seized by those who continuously develop their levels of expertise and learning methods.

If we want to be among those who direct the currents of change, we must accept learning as a life-long process, in which our patterns of basic assumptions – i.e. paradigms, play a very important role. Such paradigms change constantly during interaction with the changes taking place in an environment, so that the contemporary paradigm of a successful post-modern society and organization differs significantly from its precursors.

In the following study, we present a model of the fundamentals of the new paradigm for quality organizational change management, realized on the basis of research conducted on various contemporary approaches and models for organizational change management (ISO 9001 Standard, EFQM Excellence Model, 20 Keys, BSC (Balanced Scorecard), Six Sigma and BPR (Business Process Reengineering)), as well as extensive research conducted on a sample of 90 companies.

Key words: *paradigm, change management, quality, organization system, fundamental*

1. For the Successful, Changes Represent an Opportunity, Not a Problem

The laws of nature demand that we continuously expand the limits of our capabilities – and this striving is closely connected with changes and the act of changing. Thus, continual changes at civilizational, organizational and personal levels become manifest in response to the changes taking place in the environment. In this way, change becomes a constant; and equally so, the wish for success. Success can be defined in the broadest sense as the tendency to realize expectations – which we cannot achieve, however, if we lack the capacity to effect decisive changes at civilizational, organizational and personal levels in response to the changes in our environment. This ability to manage is contingent on our capacity to perceive reality - in which, again, paradigms play a decisive role. In this context, paradigms represent a pattern of hypothetical perceivable fundamentals, approaches to and theories of change management. Since the laws of nature are also considered paradigms, the fundamental suppositions change too, so that the subsequent change of paradigms must also be regarded as a constant.

Table 1: Meaningfully abridged definitions of the key words encountered within the field of research

Paradigm:
Pattern of presupposed perceivable fundamentals, approaches and theories
Change Management:
The capacity to exert a decisive influence on changes at civilizational, organizational and personal levels in response to changes in the environment
Quality:
Continuous improvement of success, i.e. continuously exceeding the level of achievement in realizing what is expected of us
Organization System:
A community of interdependent people linked by a specific goal, intention and program and according to specific principles and characteristics, who comprise a comprehensive whole
Fundamentals:
The most important concepts and principles of a paradigm or of a pattern of presupposed perceivable fundamentals, approaches and theories

In order to shed additional light on the indicated problem, we present a summary of the key findings of Drucker - one of the most prominent researchers of the history, theory and practice of management, who devotes the better half of one of his latest books to change management and the challenges that these changes pose to managers in the 21st century (Drucker, 2001:76):

- Changes cannot be predicted – they can only be anticipated.

- Owing to the fact that change is inevitable, the themes related to overcoming resistance to change that were topical a decade ago are giving way to new ones.

- In these times of turbulent change that we live in, change has become the rule.

- Carrying out changes is a painful and risky task, which entails above all much hard work.

- If an organization – whether it be a company, university, hospital or any other establishment – wants to survive, it must even encourage change.

- In times of rapid structural changes, those directing the changes will be the only ones to survive.

- One of the main challenges of management in the 21st century is the restructuring of organizations into conductors of change – as they need to see the opportunities inherent in change.

- A conductor of change has developed an approach that enables it to constantly discover useful changes and also knows how to take advantage of these in order to increase its success.

The preliminary findings are indicative of a need to change the existing thought pattern at personal, group and civilizational levels. They also indicate that we must become aware of the fact that only a proactive approach can ensure long-term success. Changes and effecting change are becoming a constant. The dynamics of change are increasing dramatically; new knowledge is emerging, new technologies pose new challenges, new methods of communicating offer almost unlimited possibilities of synergistic effects. But, primarily, the characteristic of the new culture of innovativeness is that it constantly generates the requirement to surpass itself in the desire for a higher quality of life.

Although a great amount of change has been caused by very many different factors, in Hammer's opinion, the key generators of change in the present and in the future are as follows (Hammer, 2001:247):

- An explosion of scientific knowledge, as the percentage of scientists who work in companies is on the increase.

- Modern telecommunications infrastructure facilitates the spreading of ideas at the speed of light.

- The presence of innovative culture, which desires change, and in which the former attitude of swearing by tradition, the tried and tested, and loyalty are giving way to the demands for whatever is newest and most modern.

The scientific and technological revolution belongs to the past. Now human history is faced with the challenges delivered by the information society. New service activities are gaining ground. That is why tourism activities (Ovsenik, 2003:392) are also increasingly being studied with growing interest in terms of the expected challenges, opportunities and problems.

We are all aware of the fact that changes are a force to be reckoned with and that they must also be taken advantage of as opportunities by means of a proactive approach, yet the various theories, approaches and models applied to seeking the answer to the question of "How" differ considerably. While the unencumbered individual sees no harm in this, from the standpoint of civilization, organizational systems and the proactive individual, however, this fact poses a great problem - or a fresh and untapped opportunity - as there must be certain constants hiding beneath the surface of all this diversity.

2.   The Paradigm as a Model of Our Perception of Reality

Drucker (2001:14) states the opinion that the term paradigm means a collection of basic suppositions on the nature of reality. In this way, paradigms take root in the subconscious of the researchers studying individual areas and thus, to a great extent, determine how a given science envisions reality. A meaningful interconnectedness of suppositions forms a paradigm, and this paradigm also has a recursive function, by means of which it directs the attention of each science to discern between what are deemed important and unimportant areas.

A Paradigm therefore does not represent reality. It is merely an attempt to mirror reality as objectively as possible. Covey (1994:21) is of the opinion that a paradigm can encompass a model, theory, perception, supposition or a system of observation. In the paradigm, he sees the way in which the individual "sees" the world within the framework of their perception, understanding and interpretation of it. He picturesquely compares the paradigm with a map, and therefore, just as a map is not the territory, so the paradigm is not the world, but a model for perceiving its reality.

The concept of human identity is of great significance in exploring our perception of reality, which to this day has still not been researched enough (Ovsenik, Ovsenik, 2002:665). In their work, the authors establish that because of the non-identical characteristics of successive states in their environment, every individual must continuously delineate the shape of his own identity. The environment, with all its multi-layered attributes of change is constantly exerting a strong influence on a person's identity, demanding of the individual a continuous proactive, self-critical self-evaluation of the realization of the image of his own personality. During this process of self-assessment as a voluntary response to the challenges and/or opportunities presented by the environment, the individual continuously develops his/her identity. The concept of identity is linked closely with the concept of a paradigm, as both tie into the personal fundamentals i.e. the truths of our reality. The credibility of an individual is assessed, in our opinion, by the extent to which his (projected) identity and his personal life paradigms match.

The systematic study of possible scenarios for the future development of societies and organizations (Vila, 2001:53) indicates that we are most likely to enter into a post-modern society and organization characterized by the principles of fragmentation, deconstruction, globalization, individualization and the new post-modern culture. The same author also states that the post-modern organization will have to be flexible, process-oriented and totally customer-oriented, while at the same time it will have to foster team-work and constantly conduct comparisons between itself and the best in its field.

Contemporary researchers present an even more systematic and holistic study of the post-modern organization with respect to the new doctrine of organization, management and organizational behavior (Ovsenik, 1999:106). The new doctrine, from this viewpoint, stands on seven pillars, which, taking the starting hypothesis that "an organization is a relationship between people", are developed from the following premises:

- The critical areas in organization/management are the contacts between co-workers.

- Every single person in an organization is a participant and therefore co-shapes and manages his contacts and is at the same time the subject of contacts with co-workers.

- The organizational (and/or management) problem is complex from the disciplinary point of view.

- In the phenomenon of the organization, the moment of awareness is essential.

- The moment of awareness opens the issue of the multi-layered nature of the phenomenon of the organization.

- The way in which one sees the world creates and can also effect a crucial change in both person and world alike within the cosmic-developmental self-organization.

- New events and new models dictate and also enable new (social) schemes/concepts of organization, management and organizational behavior.

The paradigm of a successful post-modern society and organization differs significantly from its precursors. With the transition of the paradigm (Ovsenik, Ambrož, 2000:26) into post-modernity, its characteristics lie in the shift from subsystems to systems, the structural dynamics, the shift from the objective world to the cognizable world; the shift from the structural to processes, and in the shift from objective truth to approximate description. What we are dealing with is basically a paradigmatical definition of the paradigm, as all paradigms to date have striven to be objective, realistic and mechanistic. Yet because the paradigm is a model of a way of perceiving reality (a map and not the territory itself), the post-modern paradigm is, of necessity, more limited and approximate. In our opinion, it can hardly be otherwise if it wants – together with its fundamentals – to be universal, timeless and successful. It must serve the individual, the organization and civilization alike in enabling them to manage change more effectively.

The paradigm of tourism (Ovsenik, 2003:392) activities has changed. In fact, completely new conceptual forms are emerging. Tourism has a more profound impact on the life of a nation and its identity and culture than other activities, carving the landscape to such an extent that a co-natural development cannot be assured.

In the face of the fact everything is constantly in a process of change that in our current environment, we need to give man/woman as the key agent in the struggle for the unending successful management of change a certain constant, which will serve him/her in his/her efforts for the timeless management of change. This constant must make it possible for him/her to unceasingly and proactively recognize change as opportunities and to effect the successful management of such opportunities. This constant can be recognized in the form of the new paradigm for the management of change.

3. The Theoretical Starting Points for Recognizing the Answers to the Questions: What to Change and How to Change It

Both the beginning and the end of change originate in the environment. In this day and age, changes take place within trends, the progression of which is far faster than our capacity both to keep abreast of them and to change our organizations correspondingly. In this progression, the process of learning features as the key limiting factor. It is important that we are aware of the fact that just as change and changing take place at personal, group and systemic levels, so does the process of learning.

Managing change has been the indirect and the direct theme of studies conducted by numerous researchers. In his most recent work, Hammer (2001:247), the author of BPR – Business Process Reengineering, presents his visionary view of a future, in which, on the basis of recognizable mega trends, he puts forward 9 fundamental business concepts, which are in his opinion the fundamental workings of all the excellent organizations operating in today's turbulent environment (Hammer, 2001:223-225):

- Run your business for your customers - become ETDBW (Easy-To-Do-Business-With).

- Give your customers what they really want – deliver MVA (More-Value-Added).

- Put processes first – make high performance possible.

- Create order where chaos reigns – systematize creativity.

- Measure like you mean it – make measuring part of managing, not accounting.

- Manage without structure – profit from the power of ambiguity.

- Focus on the final customer – turn distribution chains into distribution communities.

- Knock down your outer walls – collaborate wherever you can.

- Extend your enterprise – integrate virtually, not vertically.

The key words that he places emphasis on are: focusing on the customer, the process-oriented approach, management based on the facts, unleashing the creative capacities of the employees, and developing partnerships. In his opinion, the coming decades will see a supremacy of those companies on the market that will be capable of implementing these key points in the appropriate way in their everyday business operations. Yet with the above-listed nine fundamentals, Hammer only gives an answer to the question of "What to do?" while at the same time, he concludes that for the realization of the above-mentioned approach, the question of "How to do it?" is of key importance. He recommends gaining a thorough and in-depth understanding of the nine fundamentals and the implementation of the following six steps (Hammer, 2001:229-242):

- Integrate and focus your efforts.

- Give more attention than you think is needed to people issues.

- Manage different constituencies differently.

- Display committed executive leadership.

- Communicate effectively.

- Deploy in a series of steps.

The order in which these six steps are listed represents a process in which management plays the decisive role, together with its personal integrity, and communication abilities.

It would be interesting at this point also to take a look at some of the experiences of failed attempts at managing change, which function as an incentive for research and for applying the most quality-oriented approach possible. It is a well-recognized fact that thorough organizational changes cannot be implemented without a change in organizational culture – whereby in the opinion of many experts note must be taken of the following seven key steps (Hesselbein, Johnston, 2002:3):

- Scanning the environment for the two or three trends that will have the greatest impact on the organization in the future.

- Determining the implications of those trends for the organization.

- Revisiting the mission and examining our purpose and refining it until it is a short, powerful, compelling statement of why we do what we do.

- Banning the old hierarchy we all inherited and building flexible, fluid management structures and systems that unleash the energies and spirits of our people.

- Challenging the "way we have always done it" by questioning every policy, practice, procedure, and assumption, abandoning those that are of little use today or in the future – and keeping only those that reflect the desired future.

- Communicating with the few powerful, compelling messages that mobilize people around missions, goals, and values – not with 50 messages that our people have trouble remembering.

- Dispersing the responsibilities of leadership across the organization, so that we have not one leader, but many leaders at every level of the enterprise.

In these statements we find that emphasis is placed on focusing on the future, revisiting the mission, unleashing the creative capacities of the employees, the critical examination of processes, clear communications and the culture of empowerment. The content corresponds with our assertion that, in order to achieve constant success, the management of an organization must foster values that will enable the kind of communication to be established that ensures that the employees have an unambiguous and clear understanding of their duties, while at the same time imbuing them with such a motivational charge that creativity and innovativeness become an inherent part of every task they perform. This also confirms our reflections that the successful implementation of change depends upon the active participation of the leaders, their ability to lead by example (leadership), and the continuous verification of their own credibility.

In Hammer's opinion, the timely recognition of the future is of key importance, whereby he recommends taking into account the following guidelines (Hammer, 2001:250-262):

- Create an early warning system, which enables you to continuously keep track of and recognize changes, which you can then respond to quickly. Here the following must also be taken into account:

  o Develop deep insight into your customers.

  o Analyze potential as well as existing competitors.

  o Look for the seeds of the future in the present.

- Develop the ability to rapidly conceive and implement new methods of work in response to external changes.

- Develop your infrastructure to offer support to the previous two points.

Hammer emphasizes the importance of a proactive view of the future and the ability of the entire organization to adapt rapidly, whereby the appropriate infrastructure that facilitates keeping abreast of and managing opportunities and problems also plays an important part.

In managing organizational changes it is extremely important that we have a thorough knowledge and understanding of the organizational system, as this is the only way to manage change comprehensively and successfully. In this respect, models – which we can view as conceptual depictions of organizations – can be of great assistance to us. Just as paradigms are not reality, but only a mirror of it, so these models can also be regarded as mirror images of an organization. A model is only as good as the extent to which it corresponds with the actual state. An organizational model can help (Burke, 2002:177-178):

- Order and categorize the whole.
- Increase our understanding of the organization.
- Ensure a unified organizational culture.
- Interpret data on the organization for decision-making purposes.
- Direct the activities involved in managing change.

Various models developed by different authors all portray the same reality in their own different ways. Based on a critical comparison of the models developed by Weisbord, Nadler-Tushman and Tichy, Burke (Burke, 2002:192) found that these models had more commonalities than differences. All models included in this comparison were based on an openness of the organizational system and the conversion of inputs from the environment into outputs via transformational processes. These findings were critically upgraded by Burke and Litwin. Figure 1 depicts their comprehensive model of organizational change management.

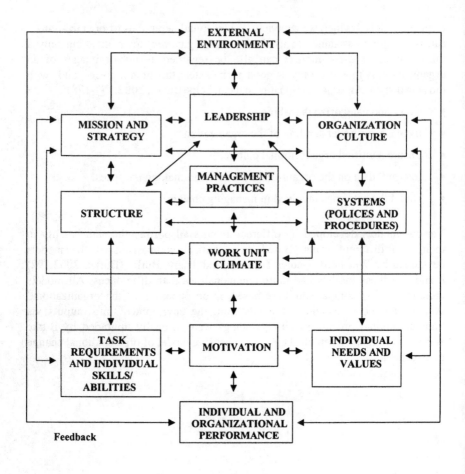

Figure1: The Burke–Litwin model of organizational change management (Burke, 2002:199)

The model with its twelve key elements interconnected along the principles of cause and effect indicates the fundamental areas within the process of managing organizational change.

Special emphasis is placed on the transformative factors, which are of key import in managing transformational changes. Transformational or radical changes are a radical response to revolutionary changes in the environment and actually represent a thorough transformation of the implemented approaches. In this model, the outer environment, the mission and strategy, leadership and organizational culture are of key importance in the achievement of individual and organizational results. The transactional factors depicted in the lower half of the model are of key importance in managing transactional changes. Transactional or gradual changes represent the constant response to evolutionary changes within the environment and actually denote a continuous improvement of the implemented procedures. The key transactional factors are the management practices, the structures, the system and its support mechanisms, the organizational climate, the demands set by the requirements, and the capabilities, needs and values of the individual.

The depicted model shows the approach taken by the process of change management. It presents an answer to the question of "How to carry out the changes," whereby it acknowledges that the leadership is in a position to be the decisive factor in the process of change management.

Again, emphasis is placed on the clear distinction between the content of the organizational changes and the process of implementing them. In the opinion of Burke (2002:14) this distinction must be clearly defined. The content of a change gives us the answer to the question of "What to change" and is closely connected with the visions and the global trends of the current environment, while the process must provide an answer to the question of "How to carry out the changes", which is connected with the application of the content.

The content of the change is obviously contingent on the current conditions in the environment and the specifics of the vision of each individual organization. An analysis of a cross-section of the current contexts and trends in development, approaches and most effective practices of current successful organizations would, in our opinion, yield an interesting array of starting points for a study of the holistic answer to the question of "What to change". This background of holistic thinking, which coincides with a specific point in time in the development of our civilization, could be seen as a paradigm, which we would classify as "the paradigm of quality change management", as it would figure as a contextual guideline for the process of changing within a given moment in civilization. Yet although it would determine the guidelines for the answer to the question: "What to change", because it is tied to the process of change itself, and therefore the answer to the question "How to change", we feel that the given verbal construct makes sense.

4. The Fundamentals of the New Paradigm for Quality Change Management

4.1 Research Description

In November 2002, extensive research was conducted on a sample of 90 organizations using the survey method. Our aim was to gain as many opinions as possible on a number of statements, which are contextually related to the approaches and practices implemented in change management. The sample encompassed both profit and non-profit organizational systems in the territory of Slovenia, which were chosen randomly from a sample of the most successful organizations over the past few years (i.e. organizations that had been nominated for the national prize for excellence in the field of business, that had gained certificates of quality, that were members of the SZK (Slovene Association for Quality), or the NFPO (National Foundation for Business Excellence). We were interested primarily in the practices, approaches and what models were implemented in the area of change management. A letter was addressed to the general manager (CEO) in each of these organizational systems, which contained five survey questionnaires and a cover letter, in which we kindly asked the CEO to distribute the remaining questionnaires among his/her fellow members of the executive team and other managers. We were interested primarily in the opinions of the organizations' leaders.

The survey questionnaire guaranteed anonymity and its content included the following segments:

Data on the respondent (his/her position in the organization, specific competencies, gender, age, level of education)

- Data on the organizational system which was the subject of the survey:
  - Profit or non-profit organizational system (industry, services, healthcare, education, state administration…)
  - Number of employees
  - Which phase of the life cycle it is currently in (birth-growth-maturity-decline)
  - Knowledge of various models:
    - ISO 9001 Standard,
    - EFQM Excellence Model,
    - 20 Keys,

- BSC (Balanced Scorecard),
- Six Sigma
- BPR (Business Process Reengineering)
  o Perception of temporal succession of the introduction of various models,
- Part A of the survey – recognizing the fundamentals of managing organizational change, with key instructions for the respondent:
  o In my opinion the approach taken in our company in managing organizational change is actually based (current state) or should be based (desired state) on the following fundamentals – (list of 8 fundamentals)
  o The higher the mark given in the range from 1 to 5 the more the respondent was considered to be in agreement with the statement
- Part B of the survey – The applicability of individual assertions in relation to the organizational system which was the subject of the survey, with key instructions for the respondent:
  o In my opinion, the following statement actually applies (current state) or ought to apply (desired state) to our organization – (list of 27 statements)
  o The higher the mark given in the range of from 1 to 5, the more the respondent was considered to be in agreement with the statement

The results will be statistically processed using the SPSS 10 program package, and we intend to carry out the final processing and interpretation of these results over the next few months. The results of the analysis will be used to shape proposals for improving the approaches taken in introducing organizational changes and to verify the following hypotheses of our research:

Hypothesis 1:	The new paradigm for quality organizational change management, recognized on the basis of the fundamentals that all the different models for change management have in common, corresponds most comprehensively with the fundamentals of the EFQM Excellence Model.
Hypothesis 2:	The choice of the specific model for change management in a concrete organizational system is contingent on the period of the life cycle it is in at the time.
Hypothesis 3:	There is a significant correlation between the quality of the leadership and how successful a company is in managing organizational change.
Hypothesis 4:	Human Resource Management (HRM) is a key process in managing organizational change.
Hypothesis 5:	Success and efficiency in organizational change management can only be achieved using the in-in approach, which calls for a balanced management of the transformational (radical) and the transactional (gradual) changes.
Hypothesis 6:	Quality organizational change management cannot be achieved without an internalization at personal level of the fundamentals of the new paradigm of quality change management.

A description of the characteristics of the sample of respondents, statistically evaluated using the SPSS 10 program package, can be found in Table 2, whereby the key commentary is as follows:

- Of the 453 questionnaires that were sent out, 263 (58%) were returned, and 244 (54%) were used for research purposes, as 19 (4%) of all the questionnaires that were returned to us were eliminated because they were incomplete.

- The target group of respondents was achieved, as 68.4 % of the respondents were general managers (CEOs) and executive team members.

- The average age of the respondents was 41 years, and male respondents were predominant (60.7%).

- The average level of education was very high, as the majority of respondents held university degrees (4-year course or more) (49.2%), followed by the group with postgraduate degrees (19.7%) and the last two groups of almost similar size with university degrees (2-year course) (16%) and secondary school education (15.1%).

Table 2: Description of the characteristics of a sample of respondents

Characteristics of the respondents (total = 244)		Frequency	Percent %
Position within the organizational unit	General manager (CEO)	37	15.2
	Member of the executive team	130	53.3
	Other managers	77	31.6
Gender	Female	96	39.3
	Male	148	60.7
Age in years (average: 41)	24 to 34	65	26.6
	35 to 44	89	36.5
	45 to 54	74	30.3
	over 55	16	6.6
Level of education	Secondary school	37	15.1
	University degree (2-year course)	39	16.0
	University degree (4-year course or more)	120	49.2
	Postgraduate	48	19.7
Type of organizational system	Non-profit - healthcare	21	8.6
	Non-profit - education	17	7.0
	Non-profit- state administration	33	13.5
	Profit - industrial organization	129	52.9
	Profit – services organization	44	18.0
Number of employees (average: 704)	up to 60	43	17.6
	61 to 120	30	12.3
	121 to 400	77	31.6
	401 to 1500	71	29.1
	over 1501	23	9.1
Stage in the life cycle that the organization is currently in	Birth	6	2.5
	Growth	104	42.6
	Maturity	111	45.5
	Decline	23	9.4

- The sample encompassed 70.9% profit organizations and 29.1% non-profit organizations. The largest group was that of profit organizations in the industrial sector (52.9%), followed by the profit services group (18%), the sample from the state administration sector (13.5%), healthcare (8.6%) and tertiary education (7%).

- The organizations had 704 employees on average. The largest group of organizations had up to 400 employees (61.7%) and the smallest group of organizations had up to 500 employees (9.1%).

- In the responses to the question of which phase of the life cycle the organization that was the subject of the survey was in, the phases of maturity (45.5%) and of growth (42.6%) occurred most frequently, while the group that recognizes the phase of decline was also of a considerable size (9.4%).

4.2   Comparative Analysis of Various Models for Organizational Change Management

In devising a new paradigm for change management we believe it is sensible to seek for the fundamentals among the holistic approaches, which have, over the last decade, empirically proven themselves through their success and innovativeness and therefore belong to the post-modern period of the development of organizational science. We believe that the starting points mentioned in our research can be used in the conceptualization of the fundamentals of the new paradigm for changes at organizational level.

In the introductory part of this study, based on the analysis of some of the best practices and approaches implemented by successful companies, as well as on the indications of upcoming trends for future development, a number of models were identified, which are implemented by the best organizations in achieving a continuous increase in their rate of success. In doing so, we focused primarily on the development and use of models in the automobile industry, as well as on the practices implemented by the finalists of the annual European Quality Award, which are presented to the public at the annual Winners' Conference and in the winners' application documents.

The models we deemed relevant were the following:

- MODEL A: The EFQM Excellence Model (source: EFQM, 1999)
  - o   A widespread model in Europe for encouraging continuous improvement, based on learning and innovativeness, designed by the European Foundation for Quality Management. It encompasses nine criteria, divided in a balanced way into "enablers" and "results".

- MODEL B: IS0 9001 Standard (source: ISO 9001, 2000)

- o A global international standard, which determines a system of quality management, based on a process-oriented approach and continuous improvement.

- MODEL C: 20 Keys to Workplace Improvement (source: Kobayashi, 1995:20)

  - o A comprehensive system for carrying out continuous improvement, developed by Professor Iwao Kobayashi, presents twenty interdependently connected tools or keys, focused primarily on the development of the production process.

- MODEL D: Balanced Scorecard (source: Kaplan& Norton, 1996:15)

  - o A comprehensive organization management model designed by Robert S. Kaplan, on the basis of a balanced collection of goals, derived from the vision, for the purpose of gauging and managing the business strategy.

- MODEL E: Six Sigma (source: Harry & Schroeder, 2000:120)

  - o A system of continuous improvement developed by Bill Smith at Motorola Company, it presents a series of programmable applied independent tools, focused mainly on decreasing the number of mistakes and on production development.

- MODEL F: Business Process Reengineering (source: Hammer, 2001:247)

  - o A process-oriented and radically transformative approach, defined by Michael Hammer, which focuses primarily on creating new values for the customer and ensuring customer satisfaction.

Already the short description of the above presented models reveals that they are similar, as the repetition of key words is evident. As we are interested in the fundamental structures of the individual models, we have depicted the fundamentals of each individual model in Table 3 in greater detail. We studied the similarities between the models as well as the contents that coincided. Concurrence among all the models was checked using a referential model, which we discerned as being the EFQM Excellence Model. The latter was chosen as referential due to the fact that it is an integral business model, which already serves in many companies as the fundamental and initial concept for the development of a business model designed to cover their specific needs. The conceptual design of the Excellence Model in essence already facilitates a meaningful upgrading of the business model with all the heretofore known models, standards and tools.

Key commentary to the comparative analysis of the various models for organizational change management (Table 3):

- The referential model with its fundamentals has proven itself as appropriate as has made it possible to check for correspondence between all the models encompassed in the comparison.

- The correspondence check was carried out based on a study of the literature and on practical examples of the individual models, as well as on the basis of the personal experience of the author of this contribution/article.

- The greatest frequency of correspondence by far was recorded in the case of the fundamental:

  o   4 - Management by processes & facts

- A high but balanced frequency of correspondence was recorded for the fundamentals:

  o   1 - Customer focus

  o   8 - Result orientation

  o   3 - People development & involvement

  o   5 - Continuous learning, innovation & improvement

  o   2 - Partnership development

  o   6 - Leadership & constancy of purpose

- By far the lowest frequency of correspondence occurred with the fundamental:

  o   7 - Public responsibility

- The differing frequency of correspondence is a result of the differences between the models, as models C and E focus more on the production process, while the other models are more system-oriented.

- The low frequency of correspondence in the case of public responsibility can also be ascribed to different civilizational cultures in which the various models developed. In Europe, i.e. within the European cultural domain, efforts are made to upgrade public responsibility, while in the American cultural domain (models E and F) public responsibility is not particularly emphasized.

- Model B (ISO 9001 standard) with its 2000 issue corresponds strongly in content with the referential model, which was to be expected, as both offer the possibility of designing a business systems model. Another parallel lies in the time frame of their development, as well as in the visible marked influence of the European cultural domain.

- Model C (20 Keys to Workplace Improvement) corresponds the most with respect to the process oriented approach, due to its focus on developing the production process, while staff development and involvement are also noticeable in the other balanced approaches.

- Model D (BSC - Balanced Scorecard) with its focus on mastering strategies corresponds in a very balanced way with all the fundamentals of the referential model, among which, however, management by processes and facts and result orientation are markedly in the forefront.

- Model E (Six Sigma) corresponds the most in terms of customer focus, management by processes and facts, and result orientation, due to its focus on reducing errors and product development. No correspondence was recorded between this model and the referential model in the area of public responsibility.

- Model F (BPR - Business Process Reengineering) shows a marked correspondence in the area of customer focus, while the other areas are balanced, with the exception of public responsibility, where no correspondence was recorded.

- The fundamentals of the referential model (A) are discernible as the universal conceptual background of the models encompassed in this comparison (B, C, D, E, F).

- We believe that the fundamentals of the referential model (A) mentioned in this study can be utilized in the conceptualization of the fundamentals of the new paradigm for change management at organizational level.

Table 3: A comparative analysis of various models for the management of organizational change

FUNDAMENTALS OF THE COMPARED MODEL	FUNDAMENTALS OF THE REFERENTIAL EFQM EXCELLENCE MODEL							
	1	2	3	4	5	6	7	8
Number of incidences of correspondence - Total→	19	13	18	36	14	12	5	19
MODEL A: EFQM EXCELLENCE MODEL								
Number of incidences of correspondence – Model→	1	1	1	1	1	1	1	1
Customer focus	■							
Partnership development		■						
People development & involvement			■					
Management by processes & facts				■				
Continuous learning, innovation & improvement					■			
Leadership & constancy of purpose						■		
Public responsibility							■	
Result orientation								■
MODEL B: ISO 9001								
Number of incidences of correspondence – Model →	1	1	1	3	1	1	1	2
Customer focus	■							
Leadership						■		
Involvement of people				■				
Process approach				■				
System approach to management								■
Continual improvement					■			
Factual approach to decision making				■				
Mutually beneficial supplier relationships		■					■	
MODEL C: 20 KEYS								
Number of incidences of correspondence – Model →	5	5	7	18	2	2	3	6
Cleaning and organizing				■				
Rationalizing the system goal alignment				■				
Small group activities			■					
Reducing work process				■				
Quick changeover technology				■				
Kaizen of operations	■			■				■
Zero monitoring of manufacturing				■				
Coupled manufacturing				■				
Maintaining machines			■					
Workplace discipline				■				
Quality assurance	■	■	■	■				■

Developing your supplier		■		■				
Eliminating waste			■	■				■
Empowering employees	■			■	■			
Cross training		■		■				
Production scheduling				■				
Efficiency control		■						■
Using information technology				■				
Conserving energy and materials	■						■	
Leading technology site technology								
MODEL D: BSC- BALANCED SCORECARD								
Number of incidences of correspondence – Model →	2	1	2	4	3	3	1	4
Financial aspect				■	■			
Customer aspect	■							
Internal processes aspect							■	
Learning and growth aspect	■						■	
MODEL E: SIX SIGMA								
Number of incidences of correspondence – Model →	3	2	2	3	2	2	0	3
Customer satisfaction with emphasis on quality	■							■
Cost reduction with emphasis on quality								
Increasing market segment through customer satisfaction and cost reduction		■		■				
MODEL F: BPR (Business Process Reengineering)								
Number of incidences of correspondence – Model →	5	2	2	3	3	2	0	1
Run your business for your customers - become ETDBW (Easy-To-Do-Business-With)	■							
Give your customers what they really want – deliver MVA (More-Value-Added)	■							
Put processes first – make high performance possible				■				
Create order where chaos reigns – systematize creativity			■		■			
Measure like you mean it – make measuring part of managing, not accounting				■				■
Manage without structure – profit from the power of ambiguity								
Focus on the final customer – turn distribution chains into distribution communities	■		■					
Knock down your outer walls – collaborate wherever you can				■				
Extend your enterprise – integrate virtually, not vertically	■							

4.3   Analysis of the Results of the Survey on the Fundamentals of Existing Approaches to Organizational Change

In conducting the survey among the leaders of a sample of 90 profit and non-profit organizations, the respondents were asked questions in the first part of the survey (Figure 2) about the fundamentals that they recognize in their environments in relation to change management.

In the survey, we listed eight fundamentals of the new paradigm for change management, which are defined in chapter 4.2 and asked the respondents to which extent they agree with each individual description of the statements. The higher the mark given in the range from 1 to 5, the more the respondent was considered to be in agreement with the statement. The respondents' opinions regarding the fundamentals of the actual approach (current state) and of the fundamentals of the approach, which they believed would ensure the successful change management (desired state) were gauged separately.

Table 4: Descriptive statistics of the respondents' perception of the fundamentals of organizational change management.

Fundamental building blocks of the new paradigm of quality organizational change management		Desired state ("ought to be in effect")			Actual state ("actually is in effect")			Difference des. s. - act. s.	
		mean	std. dev.	range	mean	std. dev.	range	dif.	range
A1	Customer focus	4.66	0.66	4	3.66	0.90	1	1	8
A2	Partnership development	4.72	0.55	2	3.44	0.89	2	1.28	6
A3	People development & involvement	4.68	0.49	3	3.01	0.93	8	1.67	1
A4	Management by processes & facts	4.54	0.63	7	3.12	0.92	6	1.42	4
A5	Continuous learning, innovation & improvement	4.76	0.50	1	3.29	0.91	3	1.47	3
A6	Leadership & constancy of purpose	4.64	0.62	5	3.08	0.91	7	1.56	2
A7	Public responsibility	4.41	0.77	8	3.29	0.95	4	1.12	7
A8	Result orientation	4.57	0.69	6	3.20	0.86	5	1.37	5
	Average	4.62	0.61		3.26	0.91		1.36	

Figure 2: Survey questionnaire for recognizing the fundamentals of organizational change management

In my opinion the approach taken in our company in managing organizational change <u>is actually based</u> (current state) or <u>should be based</u> (desired state) on the following fundamentals: (NOTE: A higher mark means that you are in greater agreement with the statement.) 1-I do not agree at all 2-I agree in part 3-I agree halfway 4-I agree almost entirely 5-I agree entirely											
A1	Customer focus The customer is the final arbiter of product and service quality and customer loyalty, retention and market share gain are best optimized through a clear focus on the needs of current and potential customers.										
	Is actually based	1	2	3	4	5	Should be based	1	2	3	4	5
A2	Partnership development An organization works more effectively when it has mutually beneficial relationships, built on trust, the sharing of knowledge and integration, with its partners.										
	Is actually based	1	2	3	4	5	Should be based	1	2	3	4	5
A3	People development & involvement The full potential of an organization's people is best released through shared values and a culture of trust and empowerment, which encourages the involvement of everyone.										
	Is actually based	1	2	3	4	5	Should be based	1	2	3	4	5
A4	Management by processes & facts Organizations perform more effectively when all interrelated activities are understood and systematically managed and decisions concerning current operations and planned improvements are made using reliable information that incorporates stakeholder perceptions.										
	Is actually based	1	2	3	4	5	Should be based	1	2	3	4	5
A5	Continuous learning, innovation & improvement Organizational performance is maximized when it is based on the management sharing its knowledge within a culture of continuous learning, innovation and improvement.										
	Is actually based	1	2	3	4	5	Should be based	1	2	3	4	5
A6	Leadership & constancy of purpose The behavior of an organization's leaders creates a clarity and unity of purpose within the organization and an environment in which the organization and people can excel.										
	Is actually based	1	2	3	4	5	Should be based	1	2	3	4	5
A7	Public responsibility The long-term interest of the organization and its people are best served by adopting an ethical approach and exceeding the expectations and regulations of the community at large.										
	Is actually based	1	2	3	4	5	Should be based	1	2	3	4	5

A8	Result orientation										
	Excellence is dependent upon balancing and satisfying the needs of all relevant stakeholders (this includes the people employed, customers, suppliers and society in general as well as those with financial interests in the organization).										
	Is actually based	1	2	3	4	5	Should be based	1	2	3	4	5

The results of the survey, statistically evaluated using the SPSS 10 program package, are presented in Table 4, while the key commentary accompanying the analysis of the survey results is as follows:

- There is no important connection between the results of the survey and the following characteristics of the respondent:
- Position within the organization.
- Specific responsibilities and competencies.
- Gender, age and level of education.
- Knowledge of various models of organizational change management.
- There is no significant connection between the results of the survey and the following characteristics of the organizational system:
    - Profit – non-profit organizational system (industry, services, healthcare, education, state administration…).
    - Number of employees.
    - Which phase of the life cycle it is currently in (birth-growth-maturity-decline).
- There is a significant difference between the perception of the fundamentals of the current state and of the desired state.
- In all the given fundamentals, the respondents saw a concept of an approach that can be taken for successful organizational change management, which is expressed by the high average grade of 4.62 (st. dev. 0.61). At the same time, however, the respondents assessed the approach taken in the actual state with an average grade of 3.26 (st. dev. 0.91).
- Regarding the approach to be taken in the **desired state,** the respondents ascribe:
    - The greatest importance to the fundamentals:
        - Continuous learning, innovation & improvement
        - Partnership development

- People development & involvement
- The least importance to the fundamentals:
    - Public responsibility
    - Management by processes & facts
    - Result orientation
- Regarding the approach taken in the **actual state**, the respondents recognize that:
    - Greatest emphasis was placed on the following fundamentals:
        - Customer focus
        - Partnership development
        - People development & involvement
    - The least emphasis was placed on the following fundamentals:
        - People development & involvement
        - Leadership & constancy of purpose
        - Management by processes & facts
- The biggest **difference between perception** of the fundamentals of the current state and of the desired state was recorded in the case of the following fundamentals:
    - People development & involvement
    - Leadership & constancy of purpose
    - Continuous learning, innovation & improvement
    - Management by processes & facts
- The cross-section of the group with the greatest differences between the current and the actual state and the group of fundamentals considered of the greatest importance for the desired state, revealed two basic opportunities for improvement:
    - People development & involvement
    - Continuous learning, innovation & improvement
- Based on the statistical tests conducted on the results of the survey and the high level of agreement recorded amongst the respondents (average grade of 4.62; st. dev. 0.61), we believe that the respondents saw the concept of an approach to the successful management of organizational change in all the given fundamentals and that the fundamentals of referential model A (the

EFQM Excellence Model) can be used in the conceptualization of the fundamentals of the new paradigm for change management.

4.4 Proposed Model of the New Paradigm Fundamentals for Quality Organization Change Management

Based on the comparative analysis of various models, the study of the literature and practical examples, as well as our survey of the fundamentals of the existing approaches to organizational change management, we designed a proposed model of the new paradigm fundamentals for quality change management.

The Burke – Litwin model for organizational change management, comprised of twelve interdependently connected variables divided evenly into transactional and transformational factors were taken as the foundation (Figure 1). The author himself, however, acknowledges that this model only supplies an answer to the question of "How to change" and points out the need for an ongoing and meaningful definition of its background (currently contingent on civilization), which must define the answer to the question of the content of the changes.

We attempted to incorporate the Burke-Litwin model in a meaningful way into the paradigmatical concept of the content paradigm of change management. In chapter 4.2, we established the following fundamentals of the new paradigm for change management at organizational level:

- Customer focus
- Partnership development
- People development & involvement
- Management by processes & facts
- Continuous learning, innovation & improvement
- Leadership & constancy of purpose
- Public responsibility
- Result orientation

A schematic diagram of the comprehensive model for organizational change management, which gives an integral summary of the findings of Burke-Litwin and our research, can be found in Figure 3.

The key commentary accompanying the proposed model:

- Burke-Litwin's model presents an answer to the question of "How to change" and gives a description of the process of change management, which has more inherent timelessness than the context itself.

- Our model of a quality paradigm provides an answer to the question of "What to change" and represents a contextual orientation for the process of change at a given civilizational moment.

- The link between the two models can be recognized in the feedback loop of the Burke-Litwin model, which links the element of external environment with the element of individual and organizational performance.

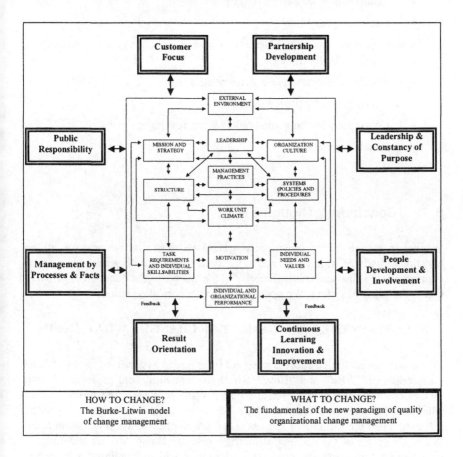

Figure 3: A model of the new paradigm for quality organizational change management

The feedback, which actually figures as a lever of continuous learning and change is influenced in a balanced way, individually and jointly, by the fundamentals of the paradigm, which supply the conceptual and guiding impulses of the contextual aspect.

- In managing the contextual aspect of transformational change it is necessary to concentrate on the following:

  - Customer focus

  - Partnership development

  - Leadership & constancy of purpose

  - Public responsibility

- In managing the contextual aspect of transactional change it is necessary to concentrate on the following:

  - People development & involvement

  - Management by processes & facts

  - Continuous learning, innovation & improvement

  - Result orientation

5. Concluding Thoughts

Whenever a problem or opportunity arises, we must find an answer to the questions, firstly of what to do, and at the same time secondly of how to do it. It is primarily the absence or insufficiency of the answer to the question of "How?" that is the cause of failure. We could set the metaphorical equation:

SUCCESS IN THE MANAGEMENT OF CHANGE = WHAT x HOW

This equation shows that success must not be regarded as a sum of factors, but as the product of "What" and "How", where the suitability and presence of both factors ensure success. Not even the best "What" will yield the desired effect when coupled with an inadequate "How", and vice versa. The essence lies in the realization that the formula of success is comprised of factors and not summands. When we speak of the successful management of change, we are speaking of balanced multiplication and not simple addition in which paradigms, approaches and models play a significant – and even a decisive role.

The process of change management with which we respond to the question of "How to carry out change?" has more inherent timelessness than the content of the changes themselves. The content, or in other words, the answer to the question of "What to change?" is characterized more by the current interaction between the organization and the environment and by the search for the contents of answers to the challenges posed by the environment – in the sense of striving to realize one's own vision and mission. We can also speak of a temporally contingent, changing pattern of assumptions, which figures as the background of thought to the efforts of mastering organizational change. This global pattern of fundamental assumptions can be recognized as a paradigm; a paradigm which can also be regarded as a momentary background - contingent on civilization - to the process of change management. Because the process of change management focuses on seeking the most effective possible answer to the question of "How to carry out the changes", i.e. the qualitative aspect, we could call the paradigm of content which is attempting to figure as the backdrop of this qualitative aspect, the "paradigm of quality change management".

References

Burke, W. W. (2002): "Organization Change: Theory and Practice", Foundations for organizational science, Thousand Oaks, California:14,177-199.

Covey, S. R. (1994): "Sedem navad zelo uspešnih ljudi" ("Seven Habits of Highly Effective People"), Založba Mladinska knjiga, Ljubljana:21.

Drucker, F. P (2001): "Managerski izzivi v 21 stoletju" ("Management Challenges for the 21st Century"), GV Založba, Ljubljana:14,76.

EFQM (1999): "The EFQM Excellence Model 1999", EFQM Publications, ISBN 90-5236-082-0, Brussels.

Hammer, M. (2001): "The Agenda: What Every Business Must Do to Dominate the Decade", Crown Business, New York:223-247.

Harry, M., Schroeder, M. (2000): "Six Sigma – The Breakthrough Management Strategy Revolutionizing the World's Top Corporations", Currency, New York:120.

Hesselbein, F., Johnston, R. (2002): "On Leading Change: A Leader to Leader Guide", Jossey-Bass, San Francisco:3.

ISO 9001 (2000): "Quality Management System", International Organization for Standardization.

Kaplan, S. R., Norton, P. D (1996): "The Balanced Scorecard – Translating Strategy into Action", Harvard Business School Press, Boston:15

Kobayashi, I. (1995): "20 Keys to Workplace Improvement", Productivity Press, Portland:20

Ovsenik, J. (1999): "Stebri nove doktrine organizacije, managementa in organizacijskega obnašanja" ("The Pillars of the New Doctrine of Organization, Management and Organizational Behavior"), Moderna organizacija, Kranj:106.

Ovsenik, J., Ovsenik, M. (2002): "Razpotja v razumevanju organizacije in managementa v razmerah globalizacije" ("Crossroads in Understanding Organization and Management in Conditions of Globalization"), Zbornik s posveta organizatorjev, FOV, (Proceedings from the Conference of Work Organizers, Faculty of Organizational Sciences), Moderna organizacija, Kranj: 661-672.

Ovsenik, M., Ambrož, M. (2000): "Ustvarjalno vodenje poslovnih procesov" ("Creative Leadership in Business Processes"), TURISTICA – College of Tourism:26.

Ovsenik, R. (2003): "Opportunities and Contradictions Involved in the Development of Tourist Destination: A model of Tourism Management in the Area of the Slovenian Alps", Organizacija in kadri, 36(6):392-399.

Vila, A. (2001): "Organizacija v postmoderni družbi" (Organization in Post-modern Society"), Moderna organizacija, Kranj:53.

Franko Milost

How to Evaluate Intellectual Capital

Abstract

Man's work is an important element of the business process, however its value is not disclosed on the assets side of the classical balance sheet. In order to show intellectual capital among assets we have to evaluate it. In this article we discuss the most significant non-monetary and monetary models of intellectual capital evaluation. Among non-financial models we discuss the Michigan, Flamholz and Ogan model. Among financial models we discuss the Replacement Costs Model, the Opportunity Costs Model, the Discounted Wages and Salaries Model and our Dynamic Model.

Key words: human resource accounting, intellectual capital, financial statements, models of intellectual capital evaluation.

1. Introduction

Man's work is an important element of the business process. The latest findings on the importance of intellectual capital for hotel companies in Slovenia are presented in Jerman (2003:400-409) and Ovsenik (2003:392-399). However, apart from its role as a means of production, products and services, its value is not disclosed on the assets side of the classical balance sheet. Are there any solid grounds for such consideration of work? Does such consideration of man's work result from underestimating the meaning of this element within the business process? And finally, isn't work (employees, human potential, intellectual capital) a factor that has a crucial influence on successful business operations? These and similar questions are within the scope of human resource accounting.

Findings on the value of intellectual capital are not new. In fact, its value has already been well recognized by pre-classical economists who treated man as an element and source of national wealth. Over time, this knowledge has increasingly matured; nowadays, however, intellectual capital is only included in financial statements in very exceptional cases.

This article identifies the most significant non-monetary and monetary models of intellectual capital evaluation. Additionally, our model of evaluation (the dynamic model) is presented.

2. Models of Intellectual Capital Evaluation

Intellectual capital may be disclosed among the assets on a balance sheet only if it is expressed in value terms. In order to disclose intellectual capital among balance sheet items, one must find a proper method for measuring its value. So far, some monetary and non-monetary models have been developed for this purpose (Brooking, 1996:42; Edvinson, 1996:13; Sawalia, 1999:33 and Edvinson, 1997:52). Some of the most important models are outlined below. Additionally, our original (dynamic), monetary model of intellectual capital valuation is presented.

2.1 Non-Monetary Models for Evaluating Employees

Among the non-monetary models, the Michigan, Flamholz and Ogan models are presented respectively below: The first two models are purely non-monetary, whilst the third one is combined, since it includes both monetary and non-monetary methods of evaluation.

2.1.1 The Michigan Model

The very first approach to a non-monetary evaluation of employees can be traced to research work conducted by the Institute for Social Research, which operates under the umbrella of the University of Michigan. The researchers of the aforementioned Institute shaped the model known as the Michigan or Likert model (named after the leading researcher of the Institute). The model defines variables that are likely to influence the effectiveness of individuals in an organization and, therefore, the successful operation of a human organization per se (Likert et al., 1969:14 ff.)

The Michigan model aims to indirectly define the value of employees in an organization. It does not enable determination of their initial value, but rather monitors value changes resulting from changes within the organizational climate. Despite the aforementioned, and though there are numerous open questions to which the authors of the Michigan model have found no suitable answers (i.e. the question of various interpretations of such results), Flamholz is of the opinion that the Michigan model represents the most successful attempt in terms of the non-monetary evaluation of employees in an organization (Flamholz, 1982:23).

2.1.2   The Flamholz Model

Contrary to Likert, Flamholz shaped his non-monetary model of human resource evaluation in terms of the individual. He wanted to explain factors that influence the value of an individual in an organization. This model consists of behavioral and economic variables.

It is based on the assumption that the value of an individual in an organization depends on two interrelated variables, namely:

a) the individual's conditional value and

b) the probability of maintaining organizational membership.

The individual's conditional value is determined as the current value of future services that may be rendered by an individual in an organization during his/her expected working life (Flamholz, 1972a: 668). Flamholz has tested his model by evaluating employees in a company registered for services in the area of accounting and business finances (Flamholz 1972b: 241-266).

2.1.3   The Ogan Model

Similar to Flamholz, Ogan shaped a model in which some of the most important variables influencing the value of an individual in an organization are defined. The model aims at evaluating human resources especially in those service enterprises where prices are not determined by the market. The prices charged for some services, for example, are determined by professional associations such as bar associations, medical associations, etc. This is a combined model since it includes both monetary and non-monetary measures. The basic idea of the model is to measure the amount of a company's long-term benefit from an employee. The value that an employee has for the company should equal the employee's long-term benefit resulting from his/her employment. This long-term benefit is determined by two factors, namely:

a) the direct benefit of an employee on account of his employment, and

b) the certainty of his employment.

The direct benefit of an employee is the sum of all expected benefits resulting from his employment. Employment certainty indicates the level of probability that the employment remains permanent. The value of an employee for the company is obtained by multiplying the values of both factors (Ogan, 1976:311).

2.2 Monetary Models of Intellectual Capital Evaluation

The following examples of monetary models of intellectual capital evaluation are presented below: the replacement costs model, the opportunity costs model, the discounted wages and salaries model, and our dynamic model.

2.2.1 The Replacement Costs Model

The replacement costs model was developed by Flamholz in 1973. The author acknowledges two concepts of replacement costs: individual and positional. Individual replacement costs are defined as a current sacrifice that is mandatory if one wants to replace an individual of particular capacity with someone (an individual) or something (a machine) of the same capacity. These costs reflect the value of an individual for a company.

However, the value of an individual largely depends on his current and future position in a company (achieved on the grounds of his capacity). The author defines positional replacement costs as those resulting from replacing the particular mandatory services of each employee in a particular work position (workplace) in a company (Flamholz, 1973:11).

The usage of this model is limited. The model requires not only an evaluation of the costs stemming from replacing an employee with someone or something, but also an evaluation of the probability that another employee (or machine) will accomplish the same work. Additionally, evaluating the costs of replacing all employees is a rather difficult task.

2.2.2 The Opportunity Costs Model

The opportunity costs model was developed by Hekimian and Jones in 1967. The basis of this model is composed of the opportunity costs of an employee – costs that reflect the value of an employee shown in case of using his alternative. Opportunity costs are defined as costs of lost benefits in a situation when an employee performs another task and/or as costs resulting from the acquisition of the necessary employee (Hekimian & Jones, 1967:108-110). According to this definition, an employee has a certain value only if he/she is an exceptional resource, namely, when his/her movement from department A to department B causes a shortage of labor force in department A. The main weakness of this model is that it does not recognize the possibility of acquiring certain work abilities by employing new people.

2.2.3 The Discounted Wages and Salaries Model

The discounted wages and salaries model was developed by Lev and Schwartz in 1971. According to this model, the value of intellectual capital is defined as the present value of anticipated (future) remuneration of employees corrected for performance ratio. The performance ratio of employees is defined as the ratio between the company's rate of return and the average rate of return in the economy. Positive correction of the present value of anticipated remuneration of employees occurs when a company's rate of return is larger than the average rate of return in the economy, and the contrary, negative correction of the present value of anticipated remuneration of employees occurs when a company's rate of return is lower than the average rate of return in the economy (Lev & Schwartz, 1971:13). Therefore, the underlying assumption is that the future value of employees' work may be evaluated by the amount of their wages and salaries.

2.2.4 The Dynamic Model for Evaluating Employees

This model is based on the economic concept of value. According to this model, the value of particular goods depends on the present and future benefits associated with these goods. This also applies to employees. Therefore, the value of employees depends on the present value of their expected future services. This definition may apply to an individual as well as to all employees within a company. Therefore, this is model is intended to evaluate

a) individual employees and

b) groups of employees (i.e. all employees within a company).

The value of individual employees may be determined directly, while the value of a group of employees may be determined indirectly, as a corrected sum of values of individual employees. This model is based on an approach usually used for evaluating the majority of tangible fixed assets by recognizing some specific features of employees.

Some may find the comparison of tangible fixed assets and employees unsuitable, morally disputable or even offensive. We apologize in advance for any misunderstandings. We treat human resources as assets not because we would like to underestimate their human characteristics, but because we would like to emphasize their economic value. This means that we treat human resources as economic goods.

2.2.4.1. Evaluating Individual Employees

As already mentioned above, this model originally aims at evaluating individual employees. We have also mentioned that the value of a group of employees may be determined indirectly, as a corrected sum of the values of individual employees. Our dynamic model for evaluating individual employees is presented in Figure 1.

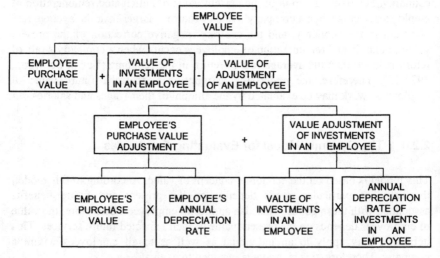

Figure 1: The Dynamic Model for Evaluating Individual Employees

Concepts and other items from the model are explained below.

2.2.4.1.1   The Purchase Value

The purchase value of tangible fixed assets normally equals the investment value associated with their acquisition. It is composed of the purchase price along with costs in relation to customs duties, transport, assembly, etc.

The purchase value of an employee is composed of investments in an employee before and directly upon his/her arrival at a company. The company does not necessary participate in all components of these investments. In the context of this evaluation model, the purchase value of an employee includes three components, namely:

a) investments in employee training,

b) investments in employee acquisition, and

c) employee opportunity costs.

Investments in employee training are associated with acquiring his/her work capacity. In this context, we talk about investment value associated with primary school, high school and university education. Investment value associated with training an employee to perform certain tasks may be defined as the usual investment needed in the process of acquiring his/her work capacity. Investment value associated with training an employee does not depend on the method of acquiring an employee. It means that the value of this investment is not subject to change when equally trained employees are of concern. An assumed value, obtained from the sum of investments needed for an employee in the process of acquiring relevant work capacity, may serve as investment value associated with employee training.

Investments in employee acquisition include:

a) investments in job advertisement and

b) investments in direct employee acquisition.

Investments in job advertisement are associated with: placing ads for an available position, interviews, evaluation of candidate suitability, etc. Investments in direct employee acquisition are associated with the medical assessment of an employee, his placement, etc.

Opportunity costs are lost benefits resulting from choosing a particular alternative. Employee opportunity costs are an individual's investments in his/her own knowledge and development. Let's assume that we are in the position of employing a university graduate. This is an individual who has successfully accomplished his/her education at all three levels: primary, secondary and tertiary.

Learning is time consuming (since it is measured in years) and tiresome. However, the results of the lasting efforts put into one's studies are not tangible material goods, but acquired knowledge and a diploma. Therefore, the decision to pursue an education forces the employee to decline remuneration that would be earned if he or she were employed during the time of study. Lost remunerations are, therefore, the opportunity costs that reflect the value of an individual's investments in his/her knowledge and development. Their value is lowest at primary school level and increases with additional years of study.

2.2.4.1.2　　　Value Adjustment

The value adjustment of a tangible fixed asset is a value of a fixed asset that is, via its usage, transferred to business effects. This value depends on the purchase value of a tangible fixed asset and its useful life. The value adjustment of an employee is a value transferred by an employee, via his cooperation in a business process, to business effects. It may be obtained by calculating the sum of an employee's purchase value adjustment and the value adjustment of investments in an employee. The calculation is presented below:

value adjustment of an employee	=	employee's purchase value adjustment	+	value adjustment of investments in an employee

The employee's purchase value adjustment is obtained by multiplying the employee's purchase value by his annual depreciation rate. The calculation is presented below:

employee's purchase value adjustment	=	employee's purchase value	x	employee's annual depreciation rate

The annual depreciation rate of an employee is obtained by dividing 1 by his useful life expressed in years. The useful life of an employee, expressed in years, is the period during which the employee renders services to the company. This period depends on the expected participation of the employee in a business process. However, there is a significant difference between a tangible fixed asset and an employee. If ownership is considered in the classical way, it quickly becomes clear that an employee is not owned, since he/she is free to leave a company. Therefore, the useful life of an employee is the period during which he or she can be reasonably expected to render services to the company. It is a period from the present to the day when an employee quits working for a company due to finding employment elsewhere, retirement or similar reasons.

The value adjustment of investments in an employee is obtained by multiplying the value of investments in an employee by the annual depreciation rate of these investments. The calculation is presented below:

value adjustment of investments in an employee	=	value of investments in an employee	x	annual depreciation rate of investments in an employee

Investments in employee acquisition include:

a) investments in direct assurance of working abilities,

b) investments in health and well-being (Macur, 1999: 50 and on) and

c) investments in loyalty to the company.

Investments in direct assurance of an employee's working abilities are those that are most profoundly relative to the employee's work in a company. They include: investments in formal and informal training and introductory training, the lower productivity of an employee during the period of his introductory training, and the lower productivity of an employee prior to his leaving the company (the opportunity costs of the company).

Investments in health and well-being are those that enable regular attendance at the workplace. They include: periodic employee medical check-ups, co-financing the lease of recreational buildings, organizing sport events and similar.

Investments in employee loyalty reduce the probability that an employee will quit working for the company due to disability, retirement or similar reasons.

The remaining item to be defined is the annual depreciation rate of investments in an employee. This rate may be obtained by dividing 1 by the useful life of investments in an employee (expressed in years) as shown below:

$$\text{annual depreciation rate of investments in an employee} = \frac{1}{\text{useful life of investments in an employee (in years)}}$$

The useful life of investments in an employee is a period during which the employee renders services to the company as a result of investments directed toward his/her employment. The duration of this period depends on the intensity of knowledge obsolescence and varies across employees. The knowledge obsolescence of employees with a technical education depends on the technical/technological development in a particular economic area while the knowledge obsolescence of a social sciences graduate depends more on scientific development in that particular area and similar.

2.2.4.1.3   The Net Carrying Amount

The net carrying amount of a tangible fixed asset is the positive difference between its purchase value and its adjusted value. It is a value that is transferred by a tangible fixed asset to business effects during its remaining useful life.

Similarly, the net carrying amount of an employee depends on two factors, namely:

a) the previously determined positive difference between the purchase value of an employee and his/her adjusted value and

b) his/her significance to a company.

The value of an employee to a company depends on his/her position in the company in terms of its organizational structure. Of course, this also influences his/her remuneration. The wages and salaries of employees are, therefore, important indicators of their value within the company. Employee wages and salaries may be defined as a factor that reflects the efficiency of the used work abilities of an individual in the company. By participating in a business process, an employee offers the company a service and receives a salary in return. The salary amount reflects the value of services offered by an individual to a company and also the employee's value to the company.

According to the above, the net carrying amount of the value of an employee must be corrected. The correction factor in this context is the ratio between the annual salary of an employee in a company and the average annual salary of an employee in a national economy. The correction factor may be defined as follows:

$$\frac{\text{annual salary of an employee in a company}}{\text{average annual salary of an employee in a national economy}}$$

2.2.4.2 Evaluating a Group of Employees

The value of a group of employees is not a simple sum of the values of individual employees – this value usually differs from such a sum due to synergetic effects.

However, a certain relationship exists between the sum of values of individual employees and the value of a group of employees. We are of the opinion that this relationship depends on the successful performance of employees in the company compared to the successful performance of employees in an entire economy.

The employees' performance coefficient serves as a measure of the successful performance of employees. It is defined as the ratio between the sum of weighted average added value per employee in a company and the entire economy over the last three years (numerator) and the sum of the number of years used (denominator). The aforementioned ratio of last year is then multiplied by a factor of 3, the ratio of two years ago by a factor of 2, and the ratio of three years ago by a factor of 1. The sum of the factors (3+2+1) equals 6. Accordingly, the performance coefficient is calculated as follows:

$$\text{employees' performance coefficient} = \frac{3 \cdot \dfrac{AA0}{BB0} + 2 \cdot \dfrac{AA1}{BB1} + \dfrac{AA2}{BB2}}{6}$$

Abbreviations in the equation mean:

AAO – added value per company employee over the last year

BBO – added value per employee in the entire economy over the last year

AA1 – added value per company employee two years ago

BB1 – added value per employee in the entire economy two years ago

The remaining two abbreviations in the equation are defined using the same logic as above.

When the value of a group of employees is to be determined, the aforementioned approach enables recognition of the overall performance of a company for a period longer than a year. When calculating, the period selection is a matter of subjective judgment; however a three-year period seems to be suitable. The business life of a company is rather intensive, and in light of this, three years would seem to be a sufficiently long period. In addition, the overall performance of a company over the last year is more accentuated than the performance of previous years.

3.   Conclusion

Employees are economic goods, and therefore we are of the opinion that their value must be known. Knowing this value is crucial for more realistic company financial statements and for anyone wishing to manage human resources efficiently. For this purpose we need an appropriate methodological framework for evaluating people, i.e. estimating the value of the intellectual capital of a company.

Two groups of models for intellectual capital evaluation exist: monetary and non-monetary models. Non-monetary models are not appropriate for disclosing intellectual capital among the assets on a balance sheet and most popular monetary models are not appropriate for general use – we can only use them in limited cases.

Accordingly, we have developed an original model for intellectual capital evaluation – the Dynamic model. This is the first step towards developing a general model for evaluating the human capital of employees.

References

Brooking, A. (1996): Intellectual capital. International Thomson Business Press, London:42.

Edvinson, L. and Sullivan, P. (1996): Developing a Model for Managing Intellectual Capital. European Management Journal, nr. 4:13.

Edvinson L. and Malone S. M. (1997): Intellectual Capital: Realizing your Company's True Value by Findings Its Hidden Roots. HarperCollins Publishers Inc., New York:52.

Flamholz, E. G. (1972a): Assessing the Validity of a Theory of Human Resource Value: A Field Study. *Empirical Research in Accounting Selected Studies, Personnel Review*:11,23,668.

Flamholz, E. G. (1972b): Toward a Theory of Human Resource Value in Formal Organizations. *The Accounting Review*:241-266.

Flamholz, E. G. (1973): Human Resource Accounting: Measuring Positional Replacement Costs. *Human Resource Management*:11.

Flamholz, E. G. (1982): *State-of-the-Art and Future Prospects*. Dickenson, Encino.

Hekimian, J. S. and Jones, C. H. (1967): Put People on Your Balance Sheet. *Harvard Business Review*:108-110.

Jerman, J. (2003): Development of human resources management in hotel enterprises: or does the system of education and training really work. *Organizacija*:400-409.

Lev, B, & Schwartz, A. (1971): On the Use of the Economic Concept of Human Capital in Financial Statements. *The Accounting Review*:13.

Likert, R. et al. (1969): How to Increase the Firm's Lead Time in Recognizing and Dealing with Problems of Managing its Human Organization. *Michigan Business Review*:14.

Macur, M. (1999): Privatization and the Quality of health-care services. Družboslovne razprave, Vol. XV, Ljubljana: 50 and on.

Milost F. (2001): *Računovodstvo človeških zmožnosti*. Fakulteta za management, Koper.

Ogan, P. (1976): A Human Resource Value Model for Professional Service Organisations. *The Accounting Review*:311.

Ovsenik, R. (2003): Opportunities and contradictions involved in the development of a tourist destination: a model of tourism management in the area of the Slovenian Alps. Organizacija:392-399.

Sawalia, B. V. (1999): Human Resource Accounting Practices in Public Undertakings in India. The Management Accountant:33.

Gordana Ivankovič

Marketing-Oriented and Strategic Management Accounting

Abstract

The author discusses the role of accounting in the hotel industry. Financial statements, which are the main output of accounting, indicate the company's success in the past, but do not provide information on the prospects for the company's operations. The author infers from this that monetary indicators that can be established on the basis of accounting figures no longer satisfy the information needs of the users. Consequently, it is necessary to design a marketing-oriented accounting model for the hotel industry, which should include the study of target groups of guests and strategic management accounting. The author believes that the role of accounting information in the hotel industry should not only comprise the compilation of data used for comparing selected companies and calculating the average in the branch in order to determine whether individual companies are better than their counterparts. The accounting department must rather prepare the data and information in a way that enables the management and other employees to focus on areas that increase the value for hotel guests and consequently the value for the owners. In addition, accounting information must enable the determination of the level of achievement in the implementation of the adopted strategy and the targets set. The paper presents the results of research conducted on Slovenian hotels' performance in the study of target groups of guests and strategic management accounting.

Key words: hotel industry, marketing-oriented accounting, strategic management accounting, market business orientation

1.   Introduction

The essential novelty in developing accounting theory is that it no longer merely involves a monetary measurement of economic categories and accounting of past business performance, but is extended to include the social responsibility of companies. Accounting is increasingly becoming a management task, and also spreading to areas outside the traditional boundaries of accounting. Such accounting is known as management accounting. This means that management accounting supplements the information obtained from cost and financial accounting with information that is otherwise not the subject of accounting monitoring. The focal characteristic of contemporary accounting is in the new awareness that accounting practice must introduce solutions that address the issue of knowledge management in the sense of measuring the intangible assets that are important. The traditional balance sheet of a company does not provide sufficient information, since it does not contain intangible assets in the sense of the concept of a knowledge-based company. Consequently, the presentation and measurement of intellectual capital is becoming increasingly important (see Gojkovič, 2001:305).

The business environment in the hotel industry is highly competitive, with each hotel directly or indirectly competing with another hotel. The highly competitive environment prompts hotel managers to meet their customers' expectations as much as possible to enable the survival and success of the business. Numerous authors state that the hotel industry does not have an appropriately developed management accounting system – MAS (Phillips, 1999:41-60; Mia and Chenhall, 1994:13; Banker et al., 1999:42) which would also provide an efficient system of measuring business efficiency. This is also proved by the following (Brander and Mcdonnell, 1995:8; Eccles, 1991:135):

- Firstly, the hotel information system lags some way behind the development in this field as it uses traditional performance measures – such as occupancy rate, profit indicators and return on investment – now thought to provide poor and misleading signals, which do not adequately support the needs of today's management.
- Secondly, the interest of management in human resources, customer satisfaction and marketing is rather apparent. However, current management information systems are relatively weak in these areas – which is perhaps not surprising, as it is well recognized that these are difficult areas to measure.
- Thirdly, it is considered essential that hotel companies develop performance measures unique to their particular market position, guest structure, management style, financial strength and competitive strategy.

- In some hotels the system of performance measurement does not enable day-to-day decision-making, thus endangering the organization's competitive advantage.

The appropriate MAS in a hotel can help managers satisfy their customers' expectations and achieve organizational goals (Damonte et al., 1997:15). Authors Mia and Chenhall (1994:9) argue that the provision of the necessary information by the MAS assists managers in enhancing the quality of the decisions they make, thereby improving their organizational performance (see also Downie, 1997:207). The MAS is expected to be available to managers in an appropriate format and on demand to satisfy managers' information needs (Dent, 1996:249; Govindarajan, 1984:128; Mia and Chenhall, 1994:12; Simons, 1990:132).

2.   Research Hypotheses

Emanuel et al. (1990:69) wrote that there was no uniform accounting system to suit all companies and conditions, which has been proven by the contingency theory. Many authors agree that the business orientation of a company is the basis for the setting up its MAS. Theory defines the structure of costs as the basis for defining business orientation. An activity with a large share of fixed costs has a market business orientation, while an activity with low fixed costs (production) is cost-oriented. The hotel industry has around 60 percent of fixed costs (Brander and Harris, 1997:113; Worldwide Hotel Industry Study, 1999). This represents a large share of fixed costs and therefore hotels must orientate themselves towards revenue management and yield management.

Downie (1995:214) points out the importance of close cooperation in the preparation of information by employees in marketing, planning and/or controlling with the employees in accounting and the departments, as well as with department managers. Similar ideas can be inferred from the work by Dunn and Brooks (1990:85) on market share profit analysis, by Quain (1992:58) on sales-mix profitability, and by Noon and Griffin (1997:78) on buyer profitability analysis. Acceptance and introduction of these theories in practical operations is, however, a long-term process.

It is characteristic of hotel companies that they operate in a very competitive environment. Competition brings threats and opportunities. A proactive company studies closely the situation in the environment, takes account of the economic, social and technological changes, and accepts opportunities as a challenge (Rolfe, 1992:35). When a hotel designs a new range of products to address this opportunity, its new business idea is frequently adopted by its competitors, which reduces the life cycle of a certain product, i.e. service. Thus hotels – marketing managers – are forced to constantly move ahead in order to retain their positions. The entry of new competitors in the hotel industry is easy, since almost every hotel manipulates its prices through special discounts and package arrangements. Management is aware that the offer must be interesting to guests; therefore, the needs of the customers are of primary importance in the decision-making process. In order for management to be successful in this, it must be familiar with the competition and know which differentiated products or services (with low costs) can form its differential advantage.

Strategic management accounting (hereinafter: SMA) focuses on adding value for customers based on the competition, market share and long-term strategic planning and/or adopted strategy (Bromwich and Bhimani, 1994:281). Collier and Gregory (1995:135) state that the use of SMA is increasing within the hotel industry, which is evident from budgeting, development of market conditions, as well as from market and competition analysis. SMA is accounting for strategic management, and as such aligns budget, implementation and business strategy controlling. Collier and Gregory (1995: 137) also believe that English hoteliers mostly use SMA for:

- preparation of information for strategy support and development and

- monitoring and control of the market, competition and their price/cost structures.

Phillips and Mountinho (1998:171) also state that the use of SMA is increasing, but its impact on the hotel industry is mostly the result of external influences by other industries. It enables hotel management to potentially solve deficiencies.

2.1   Definition of the Research Hypotheses

A hotel will be successful in the long run if it has satisfied guests who return again and again, and contribute with their visits to the hotel's achievements. The market is not a homogenous group of guests; each individual has his/her own requirements. Guests have different needs and their behavior is varied. The hotel must identify these differences, classify guests according to their target interest groups, and design its range of services for each target group of guests (Noone and Griffin, 1997:75). On this basis, individual hotels offer special discounts in order to increase promotion effects and simultaneously to maximize their occupancy rate. A significant element in a hotel's successfulness is also the adopted business strategy and strategic accounting, providing the data for monitoring business strategy implementation.

Long-term business plan preparation is heavily dependent on future predictions of two important elements: future occupancy rate and average revenue per room, connected to the costs of other activities (Collier and Gregory, 1995:135).

Two research hypotheses have been designed concerning the target group of guests, business strategy and the role played by accounting information in designing the range of services in a hotel (Ivankovič, 2004:172-173):

Hypothesis 1

The target group of guests and the length of their stay in a hotel are correlated factors, therefore, it is important to identify the target groups of guests, define their characteristics and predict the future occupancy rate of hotel rooms.

Hypothesis 2

The business performance of hotels with business strategies and strategic accounting is better. The criterion for strategy design is the calculation for the coming medium-term period (3 years or more).

2.2 Research Method

In 2002 we conducted vast research into measuring the business performance in Slovenian hotels. The questionnaire designed by the researchers to investigate the situation was sent to 51 hotels, i.e. all hotels in Slovenia with over 100 rooms. The questionnaires were filled out and returned by 39 hotels (76 % of the total number). The hotels that did not respond are located in various parts of Slovenia and are of different sizes, i.e. the unanswered questionnaires do not mean that a certain area in Slovenia or hotels of a certain size were omitted. The questionnaire was answered by the directors of hotels and heads of the accommodation, food, beverage, accounting and marketing departments. For the purpose of this paper, we are highlighting the findings of the research in terms of strategy adoption and target setting by hotels.

2.3 Criteria for Measuring the Variables

According to the hypotheses set, we also defined the dependent and independent variables. **The independent variables are**: identification of the target groups of guests and their characteristics, and the use of strategic accounting. As a **dependent variable**, we have determined the business performance of the hotel, measured by monetary and non-monetary indicators.

2.3.1 Measurement of Independent Variables

1) The identification of the target groups of guests and their characteristics was established by means of:

- assessing the significance of the target groups of guests for an individual hotel,
- monitoring the share of permanent guests and break-down by target groups of guests,
- the shortest and the longest average stay of target groups of guests in a hotel, and
- predicting the future level of room occupancy by target groups of guests.

2) A hotel has a business strategy and SMA if:
- it has adopted a long-term strategy which is used as the basis for the strategic plan referring to a period of at least three years,
- it has designed a budget balance sheet for at least three years in advance, and
- it follows certain parameters of SMA (its own share and market shares of the major competitors as well as their prices and costs).

2.3.2 Measurement of Dependent Variables

We have already mentioned that successfulness greatly depends on the right goals being set; and depending on the latter, defining the criteria for measuring business performance. It is also known that greater success is achieved by those companies that set more goals for themselves, not only profitability. The reason is that the latter is a short-term goal that can only be realized in the long run if certain conditions are fulfilled. Their fulfillment is measured by means of other, not just accounting, criteria. Due to their comparability, two indicators were used as the **monetary criteria** for measuring business performance (we took into account the period of the last five years and the figures were deflated), namely:

- **average net income from sales per room** (USALI[1] basis) and
- **average profit per room**.

According to Foster et al. (1996:13), the **non-monetary criteria for business performance** that also indicate the conditions for future successful operation are:

- **number of new products** (we have restricted ourselves to the period of the last two years) and
- **average share of permanent guests over the period of five years** (it costs as much as five times less to keep permanent guests than to attract new ones).

In the hotel industry, labor costs are the most important group of costs because of their prevailingly fixed orientation, and, as such, represent a restricting factor in internal efficiency (Čižmar and Šerič, 1999:312). Therefore, we also measured:

- **the number of employees per room** (over the period of five years).

3. Results

Our preliminary research revealed that Slovenian hotels do not monitor target groups of guests by profitability[2] and for this reason our **Hypothesis 1** claims that **hotels relate target groups of guests to the length of their stay, examine their target groups of guests and on this basis forecast their future occupancy rate**. Relating target groups of guests to the length of their stay in a hotel is statistically representative, since as many as 92.3% of respondents monitor this correlation. On average, foreign tourists are most important for the hotels (average assessment at 4.72). However, it is surprising that only one hotel out of the total number of 34 hotels responded that foreign guests stay longest at the hotel[3] (9 days). This implies a conviction or even a distorted perception that

---

[1] USALI - Uniform System of Accounts of Lodging Industry is a standard system of accounting report at the level of a unit - hotel.

[2] On the income side, the profitability of consumers includes not only consumption per boarding guest, but also additional consumption (outside full-board accommodation). In some cases the information systems in the hotels under consideration do not cover a particular guest with regard to basic hotel boarding (accommodation, food and drinks) let alone monitor individual target groups of guests with regard to additional consumption in a hotel).

[3] Information systems in Slovenian hotels do not allow for a complete monitoring of the profitability of consumers and this is the reason why we took into account the average number of days of stay per target group of guests in a hotel.

foreign tourists are "a priori" the best guests, which might stem from the situation before the war for Slovenian independence. However, **it does not confirm the correlation between the length of the stay of guests per target group and the assessment of their importance for the hotel** (Appendix 1 and 2). The foregoing considerations point to the fact that the surveyed hotels do not pay much attention to examining their target groups. The hotels stated 2 to 16 days as the longest stay of guests in a hotel, with the exception of eight spa hotels (representing 23.5% of the examined hotels that responded to the survey) with an average stay of almost eight days.

The research also confirms the fact that all the hotels in the survey predict their occupancy rate for the next year. An interesting fact is that those hotels which monitor target groups of guests (92.3%) very often generally state that the basis for forecasts is the attained occupancy rate in the past year (AM[4] 4.29).

A limited number of hotels (7.7%) which do not relate target groups of guests to the length of their stay in a hotel base their forecasts (average assessment at 3.67) on general economic growth trends on the tourism market. On the basis of the Pearson correlation the responses received only confirmed a correlation between the two target groups of guests in the hotels under consideration, namely between business travelers and participants at small seminars (of less than 100 participants), where one variable does not affect the other at the level of 64 % (Pearson coefficient being 0.80 - Appendix 2) period of stay of target groups of guests. However, there is a correlation between the importance of target groups of guests and the importance of target groups of guests registered as regular guests, which is not linear. Groups of guests who have been regular guests for a number of years present the third most important element for forecasting (AM 3.76).

The foregoing considerations infer only a partial possible confirmation of Hypothesis 1, namely:

- hotels recognize target groups of guests and relate them to the length of thei stay, and
- forecast their future occupancy rate.

The research did not confirm the definition of the features of target groups of guests, since there is no statistically representative correlation between the target groups of guests and the same being used as the basis for forecasts. Likewise, there is no correlation between the average assessment of the importance of a target group of guests and a target group of guests with the longest average stay in a hotel.

---

[4] AM – arithmetic mean of responses

Our **Hypothesis 2** claims that **hotels with a developed business strategy and SMA** (Group 1) **have better business performance than those which have no long-term strategy and strategic accounting** (Group 2). Criteria for business strategy and SMA have been established together with the criteria for measuring independent variables by way of neglecting to monitor the costing structure as one of the parameters of SMA. Horwath research on the basis of USALI, which would reflect the costing structure, is not conducted in Slovenia. Business performance has been measured on the basis of basic monetary and non-monetary indicators, which have been defined together with the criteria for measuring dependent variables.

On the basis of the analysis of the responses received, only five (5) hotels met the requirements for the first group and ten (10) were included in the second group on the basis of their complete responses. The data are given in tables 1 and 2.

Table 1: Hotels with a developed business strategy and strategic accounting (Group 1)

No	TH	S	NC/ room	FR/ room	FR/ emp	No emp/ room	SRG	NP
1.	SPA	3	3,957	108	259	0.46	0.52	0
2.	SEA	3	2,964	80	245	0.35	0.12	1
3.	O	3	1,883	-496	-1,485	0.33	0.08	4
4.	SPA	3	3,606	-106	-114	0.88	/	0
5.	SEA	5	3,753	128	106	1.19	/	0
	AM		3,461	5				

Source: Kavčič in Ivankovič, 2003:661.

Table 2: Hotels without a developed business strategy and strategic accounting (Group 2)

No	TH	S	NC/ room	PR/ room	PR/ emp	No emp/ room	SRG	NP
1.	SEA	3	3,976	327	958	0.34	/	2
2.	O	3	1,703	-146	-596	0.24	/	0
3.	O	3	3,902	549	1,298	0.43	0.05	0
4.	SPA	3	1,026	159	562	0.28	0.17	6
5.	SPA	4	1,758	223	739	0.30	0.19	6
6.	O	3	3,868	129	363	0.31	/	1
7.	O	4	3,444	-127	-257	0.47	0.03	2
8.	SEA	4	4,892	1,397	4,668	0.29	0.05	30
9.	SPA	4	10,202	1,581	1,231	1.29	0.23	30
10.	O	4	2,685	-174	-323	0.53	0.04	2
	AM		3,537	321				

Source: Kavčič in Ivankovič, 2003:662.

Key for tables 1 and 2:

No – hotel number

TH – type of hotel (SEA – seaside, SPA – spa, O – other)

S – number of stars of a particular hotel

NC/room - net income of hotel from sales per room

PR/room – profit/loss of hotel per room

PR/emp – profit/loss of hotel per employee

No emp/room - number of employees per room

SRG - share of regular guests out of all guests over the period of five years

NP - number of new products over the last two years

AM – arithmetic mean or mean value

The tables show that the average value of indicators of net income per room and financial results per room are higher in the second group of hotels, which is the group of hotels which do not have a long-term strategy and do not use SMA for their performance. The average number of employees per room greatly differs within the two groups compared and ranges for the first group from 0.33 to 1.19

employees per room and for the second group from 0.24 to 1.29 employees per room. The share of regular guests is not monitored in two out of five hotels of the first group under consideration and in three hotels of the second group. The number of new products over the last two years has been monitored only by two hotels of the first group and by eight hotels of the second group. With regard to two entries of 30 new products over the last two years we need to stress that these are versions of the same type of product and that these two entries must be considered with a certain element of caution.

On the basis of the foregoing considerations it can be inferred that **hotels with a developed business strategy and elements of SMA achieve no better performance results than those without such strategies.** This means that Hypothesis 2 does not correspond to the practice of Slovenian hotels and can thus be rejected, despite the fact that it has been confirmed by many successful tourist countries. Indeed, the issue here is how to establish and implement a strategy. It is hard to believe that what is relevant elsewhere is irrelevant for Slovenian hotels. The answers point to the conclusion that operations in the hotels under consideration are inert and carried out on the basis of ad-hoc actions rather then professional and sound decisions. This might also be one of the reasons for bad business results. As a matter of interest we would like to point out that spa and other hotels generally consider statistical information as the most important source for monitoring their competition (the AM of importance for the spa hotels is 3.94 and for the group of other hotels is 4.08), while seaside hotels rely on their annual reports (the AM of importance is 4.43%).

3.   Discussion

The results of Slovenian hotels are not able to fully confirm the set hypotheses, which is in fact quite surprising. Much research has been conducted worldwide on the topic researched here in Slovenia and with similar results, i.e. that hotels are aware of the need to structure their range of products and services for each target group of guests and that the application of SMA contributes to the business performance of a hotel. Slovenian hotels relate a target group of guests to the length of their stay and predict their future occupancy rate. However, it seems that the examination of target groups stops at this point. Hotels with developed business strategies and SMA have not demonstrated any better performance results than those without such strategies. The responses received from Slovenian hotels indicate a relatively low level of education, which is compensated by management work experience (Ivankovič, 2004:183-184). A precondition for success is not only the existence of elements which should lead to good performance results in the future but also the right combination of these.

The absence of the right combination could possible be responsible for the situation in Slovenian hotels. Kavčič and Ivankovič (2004:120) have drawn attention to possible reasons relating to the verification of the hypotheses. They maintain that Slovenian hotels do not properly define their objectives. It is not the MAS that can be blamed for the feeble correlation between MAS development and good performance, but rather the improperly defined objectives. Finally, the population of the Slovenian hotels under consideration is very heterogeneous, what makes it impossible to adopt some generally accepted conclusions.

Bibliography

Banker, R. et al. (1999): "An empirical investigation of an incentive plan based non-financial performance measures", The Accounting Review:42.

Brander, B. J. and Mcdonnell, B. (1995): "The Balanced score-card: short term guest or long-term resident?", International Journal of Contemporary Hospitality Management 7 (2/3):7-11.

Brander B. J., Harris, P. J. (1997): Organizational Culture and Control in a Strategic Planning Context: Implications for the International Hospitality Industry. Teare, R. et al. (eds.): Global Directions for Hospitality and Tourism Development. London, Cassell:105–131.

Bromwich, M., Bhimani, A. (1994): Management Accounting: Pathways to Progress. London : Chartered Institute of Management Accounts:281.

Collier P., Gregory A. (1995): Strategic management accounting: a UK hotel sector case study: International Journal of Contemporary Hospitality Management. Vol. 7, No 1:135,137.

Čižmar, S., Šerič, M. (1999): "Marketinški uspjeh i unutrašnja efikasnost hrvatske hotelske industrije", Turizam 4:300-315.

Damonte, L. et al. (1997): "Brand affiliation and property size effects on measures of performance in lodgings industry", Hospitality Research Journal 20 (3):1-16.

Dent, J. (1996): Global competition: challenges for management accounting and control, Management Accounting research 7:247-269.

Downie, N. J. (1995): The Use of Accounting Information in Hotel Marketing Decisions. In Harris, P. J., (ed.): Accounting and Finance for the International Hotel Industry. Oxford: Butterworth-Heinemann:202-221.

Downie, N. J. (1997): "The use of accounting information in hotel marketing decisions", International Journal of Hospitality Management 16 (3):305-312.

Driben, L. (1993): The service edge: Sales and Marketing Management.

Dunn, K. D., Brooks, D. E. (1990): Profit Analysis Beyond Yield Management. Cornell Hotel and Restaurant Administration Quarterly, 31 3, 80-90.

Eccles, R. (1991): "The performance measurement manifesto", Harvard business Review 69 (1):131-137.

Educational Institute of the American Hotel & Motel Association, (USALI), Uniform System of Accounts for the Lodging Industry, 9th revised ed., Hotel Association of New York City, 1996:238.

Emmanuel C. et al. (1990): Accounting for Management Control, Chapman & Hall, London:69.

Foster, G. et al. (1996): "Customer Profitability Analysis: Challenges and New Directions" Cost Management (Spring):5-17.

Fisher C., Lawrance, J. (1994): Low prices alone will not fly in biz travel. *Advertising Age.*

Gojkovič, B. (2001): Teoretične podlage za diferenciranje računovodskih rešitev. Doktorska disertacija: Univerza v Ljubljani, Ekonomska fakulteta:305.

Govindarajan, V. (1984): "Appropriateness of accounting data in performance evaluation: an empirical examination of environment uncertainty as an intervening variable", Accounting Organization and Society 9 (2):125-136.

Hotel Association Of New York City (1996): Uniform System of Accounts for the Lodging Industry, 9the.., Educational Institute of the American Hotel and Motel Association, East Lansing, MI:238.

Ivankovič, G. (2004): Računovodsko merjenje uspešnosti poslovanja v hotelirstvu, Doktorska disertacija: Univerza v Ljubljani, Ekonomska fakuleta:280 p. + Appendix 78.

Kavčič, S., Ivankovič, G. (2003): "Raziskovalno poročilo: Vzpostavljanje razmer za pridobivanje računovodskih in drugih podatkov potrebnih za metodologijo USALI v Sloveniji":133.

Kavčič, S., Ivankovič, G. (2004): Influence of management accounting on efficiency of tourism and hospitality management in Slovenia. Tourism & Hospitality industry (17-th Biennial International Congress) – Opatija:659-677.

Kotas, R. (1973): Market orientation Hotel, Catering and Institutional Management Journal, July, 5-7.

Mia, L., Chenhall, R. (1994): "The usefulness of MAS functional differentiation and management effectiveness", Accounting, Organization and Society 19 (1), 1-13.

Noone, B., Griffin, P. (1997): Enhancing yield management with customer profitability analysis. *International Journal of Contemporary Hospitality Management,* Vol. 9, No 2:75-79.

Phillips, P. A., Moutinho, L. (1998): The Marketing Planning Index (MPI): A Tool for Measuring Marketing Planning Effectiveness. Journal of Travel and Tourism Marketing, 17, (3):41–60.

Phillips, P. A. (1999): "Performance measurement system and hotels: a new conceptual framework", International Journal of Hospitality Management, Vol. 18 (No. 2):171-182.

Quain, W. J. (1992): Analyzing Sales-mix Profitability. Cornell Hotel and Restaurant Administration Quarterly, 33, (2):57–62.

Rolfe, A. J. (1992): Profitability reporting techniques bridge information gap. *The Journal of Business Strategy* 13 (1):32-37.

Simons, R. (1990): "The role of management control systems in creating competitive advantage: new perspective", Accounting, Organizations and Society 15:127-143.

Worldwide Hotel Industry Study 1998 (1999): New York : Horwath International:58.

Appendix 1: Correlation between forecast basis and the importance of target groups of guests for the hotel

Correlations

		Forecast Based on the Previous-Year Level	Forecast Based on the General Economic Trends in Tourist Market	Forecast Based on Arrivals in Particular Seasons in Several Years	Forecast Based on Arrivals in All Seasons in Several Years	Forecast Based on Arrivals According to Target Groups in Several Years	Forecast Based on Individual Arrivals in Several Years	Guest Target Group with the Longest Stay in the Hotel in 2001	The Longest Stay of the Target Group (Number of Days)
Forecast Based on the Previous-Year Level	Pearson Correlation	1	0,055	0,344	0,383	0,418	0,245	0,050	0,075
	Sig. (2-tailed)	.	0,752	0,037	0,028	0,010	0,156	0,781	0,675
	N	37	35	37	33	37	35	34	34
Forecast Based on the General Economic Trends in Tourist Market	Pearson Correlation	0,055	1	0,420	0,157	0,065	0,238	0,044	0,014
	Sig. (2-tailed)	0,752	.	0,012	0,384	0,712	0,182	0,810	0,941
	N	35	35	35	33	35	33	32	32
Forecast Based on Particular Arrivals in Several Seasons Years	Pearson Correlation	0,344	0,420	1	0,673	0,582	0,643	-0,225	-0,152
	Sig. (2-tailed)	0,037	0,012	.	0	0	0	0,201	0,391
	N	37	35	37	33	37	35	34	34
Forecast Based on Arrivals in All Seasons in Several Years	Pearson Correlation	0,383	0,157	0,673	1	0,750	0,677	-0,148	-0,221
	Sig. (2-tailed)	0,028	0,384	0	.	0	0	0,434	0,240
	N	33	33	33	33	33	32	30	30

Forecast Based on Arrivals According to Target Groups in Several Years	Pearson Correlation	0,418	0,065	0,582	0,750	1	0,649	-0,301	0,072
	Sig. (2-tailed)	0,010	0,712	0	0	.	0	0,084	0,686
	N	37	35	37	33	37	35	34	34
Forecast Based on Individual Arrivals in Several Years	Pearson Correlation	0,245	0,238	0,643	0,677	0,649	1	-0,333	-0,068
	Sig. (2-tailed)	0,156	0,182	0	0	0	.	0,063	0,713
	N	35	33	35	32	35	35	32	32
Guest Target Group with the Longest Stay in the Hotel	Pearson Correlation	0,050	0,044	-0,225	-0,148	-0,301	-0,333	1	-0,138
	Sig. (2-tailed)	0,781	0,810	0,201	0,434	0,084	0,063	.	0,438
	N	34	32	34	30	34	32	34	34
The Longest Stay of the Target Group (Number of Days)	Pearson Correlation	0,075	0,014	-0,152	-0,221	0,072	-0,068	-0,138	1
	Sig. (2-tailed)	0,675	0,941	0,391	0,240	0,686	0,713	0,438	.
	N	34	32	34	30	34	32	34	34

*. Correlation is significant at the 0.05 level (2-tailed);
**. Correlation is significant at the 0.01 level (2-tailed).

Source: Ivankovič, 2004, p. LVII

Correlations

		Forecast Based on the Level of the Previous Year	Forecast Based on General Economic Trends In Tourist Market	Forecast Based on Arrivals In Particular Seasons In Several Years	Forecast Based on Arrivals In All Seasons In Several Years	Forecast Based on Arrivals According To Target Groups In Several Years	Forecast Based on Individual Arrivals In Several Years	Guest Target Group With The Shortest Stay In the Hotel In 2001	The Shortest Stay of the Target Group (Number of Days)
Forecast Based on The Level of The Previous Year	Pearson Correlation	1	0,055	0,344	0,383	0,418	0,245	0,071	-0,112
	Sig. (2-tailed)	.	0,752	0,037	0,028	0,010	0,156	0,705	0,550
	N	37	35	37	33	37	35	31	31
Forecast Based on General Economic Trends in Tourist Market	Pearson Correlation	0,055	1	0,420	0,157	0,065	0,238	0,144	-0,090
	Sig. (2-tailed)	0,752	.	0,012	0,384	0,712	0,182	0,455	0,642
	N	35	35	35	33	35	33	29	29
Forecast Based on Arrivals in Particular Seasons in Several Years	Pearson Correlation	0,344	0,420	1	0,673	0,582	0,643	0,152	-0,231
	Sig. (2-tailed)	0,037	0,012	.	0	0	0	0,416	0,211
	N	37	35	37	33	37	35	31	31
Forecast Based on Arrivals in All Seasons in Several Years	Pearson Correlation	0,383	0,157	0,673	1	0,750	0,677	0,165	0,147
	Sig. (2-tailed)	0,028	0,384	0	.	0	0	0,402	0,457
	N	33	33	33	33	33	32	28	28

Forecast Based on Arrivals According to Target Groups in Several Years	Pearson Correlation	0,418	0,065	0,582	0,750	1	0,649	-0,053	0,127
	Sig. (2-tailed)	0,010	0,712	0	0	.	0	0,777	0,496
	N	37	35	37	33	37	35	31	31
Forecast Based on Individual Arrivals in Several Years	Pearson Correlation	0,245	0,238	0,643	0,677	0,649	1	-0,058	-0,209
	Sig. (2-tailed)	0,156	0,182	0	0	0	.	0,760	0,269
	N	35	33	35	32	35	35	30	30
Guest Target Group with The Shortest Stay in the Hotel	Pearson Correlation	0,071	0,144	0,152	0,165	-0,053	-0,058	1	0,073
	Sig. (2-tailed)	0,705	0,455	0,416	0,402	0,777	0,760	.	0,696
	N	31	29	31	28	31	30	31	31
The Shortest Stay of the Target Group (Number of Days)	Pearson Correlation	-0,112	-0,090	-0,231	0,147	0,127	-0,209	0,073	1
	Sig. (2-tailed)	0,550	0,642	0,211	0,457	0,496	0,269	0,696	.
	N	31	29	31	28	31	30	31	31

. Correlation is significant at the 0.05 level (2-tailed).
.. Correlation is significant at the 0.01 level (2-tailed).

Source: Ivankovič, 2004, p. LVIII

Appendix 2: Correlation between target groups of guests in the researched hotels

Correlations

		The Importance of Executives for the Hotel	The Importance of Domestic Tourists for the Hotel	The Importance of Foreign Tourists for the Hotel	The Importance of Domestic Groups Of Tourists for the Hotel	The Importance of Foreign Groups of Tourists for the Hotel	The Importance of Seminars with Less Than 100 Participants for the Hotel	The Importance of Seminars With More Than 100 Participants for the Hotel	The Importance of Health Resort Guests for the Hotel	Guest Target Group with the Longest Stay in the Hotel in 2001	The Longest Stay of the Target Group (Number of Days)
The Importance of Executives for the Hotel	Pearson Correlation	1	-0,055	0,349	0,024	0,180	0,800	0,455	-0,476	-0,317	-0,300
	Sig. (2-tailed)	.	0,756	0,043	0,897	0,317	0	0,010	0,007	0,083	0,102
	N	34	34	34	32	33	33	31	31	31	31
The Importance of Domestic Tourists for the Hotel	Pearson Correlation	-0,055	1	0,033	0,656	0,290	-0,076	-0,192	0,528	-0,076	0,190
	Sig. (2-tailed)	0,756	.	0,849	0	0,091	0,666	0,285	0,002	0,669	0,282
	N	34	37	36	35	35	35	33	33	34	34
The Importance of Foreign Tourists for the Hotel	Pearson Correlation	0,349	0,033	1	0,157	0,322	0,324	0,042	0,271	-0,216	0,097
	Sig. (2-tailed)	0,043	0,849	.	0,374	0,060	0,058	0,818	0,127	0,228	0,593
	N	34	36	36	34	35	35	33	33	33	33
The Importance of Domestic Groups of Tourists for the Hotel	Pearson Correlation	0,024	0,656	0,157	1	0,518	0,093	-0,082	0,316	0,171	0,105
	Sig. (2-tailed)	0,897	0	0,374	.	0,002	0,600	0,649	0,078	0,350	0,566
	N	32	35	34	35	34	34	33	32	32	32
The Importance of Foreign Groups of Tourists for the Hotel	Pearson Correlation	0,180	0,290	0,322	0,518	1	0,119	-0,010	0,073	0,077	0,111
	Sig. (2-tailed)	0,317	0,091	0,060	0,002	.	0,502	0,956	0,691	0,676	0,546
	N	33	35	35	34	35	34	33	32	32	32

The Importance of Seminars with Less than 100 Participants for the Hotel	Pearson Correlation	0,800	-0,076	0,324	0,093	0,119	1	0,662	-0,382	-0,249	-0,185
	Sig. (2-tailed)	0	0,666	0,058	0,600	0,502	.	0	0,028	0,169	0,310
	N	33	35	35	34	34	35	33	33	32	32
The Importance of Seminars with More than 100 Participants for the Hotel	Pearson Correlation	0,455	-0,192	0,042	-0,082	-0,010	0,662	1	-0,250	-0,198	-0,296
	Sig. (2-tailed)	0,010	0,285	0,818	0,649	0,956	0	.	0,167	0,295	0,113
	N	31	33	33	33	33	33	33	32	30	30
The Importance of Health Resort Guests for the Hotel	Pearson Correlation	-0,476	0,528	0,271	0,316	0,073	-0,382	-0,250	1	-0,127	0,487
	Sig. (2-tailed)	0,007	0,002	0,127	0,078	0,691	0,028	0,167	.	0,502	0,006
	N	31	33	33	32	32	33	32	33	30	30
Guest Target Group with the Longest Stay in the Hotel In 2001	Pearson Correlation	-0,317	-0,076	-0,216	0,171	0,077	-0,249	-0,198	-0,127	1	-0,138
	Sig. (2-tailed)	0,083	0,669	0,228	0,350	0,676	0,169	0,295	0,502	.	0,438
	N	31	34	33	32	32	32	30	30	34	34
The Longest Stay of the Target Group (Number of Days)	Pearson Correlation	-0,300	0,190	0,097	0,105	0,111	-0,185	-0,296	0,487	-0,138	1
	Sig. (2-tailed)	0,102	0,282	0,593	0,566	0,546	0,310	0,113	0,006	0,438	.
	N	31	34	33	32	32	32	30	30	34	34

. Correlation is significant at the 0.05 level (2-tailed).
". Correlation is significant at the 0.01 level (2-tailed).
Source: Ivankovič, 2004, p. LIX

Correlations

		The Importance of Executives for the Hotel	The Importance of Domestic Tourists for the Hotel	The Importance of Foreign Tourists for the Hotel	The Importance of Domestic Groups of Tourists for the Hotel	The Importance of Foreign Groups of Tourists for the Hotel	The Importance of Seminars with Less than 100 Participants for the Hotel	The Importance of Seminars with More than 100 Participants for the Hotel	The Importance of Health Resort Guests for the Hotel	Guest Target Group with the Shortest Stay in the Hotel in 2001	The Shortest Stay of the Target Group (Number of Days)
The Importance of Executives for the Hotel	Pearson Correlation	1	-0,055	0,349	0,024	0,180	0,800	0,455	-0,476	-0,259	-0,007
	Sig. (2-tailed)	.	0,756	0,043	0,897	0,317	0	0,010	0,007	0,175	0,972
	N	34	34	34	32	33	33	31	31	29	29
The Importance of Domestic Tourists for the Hotel	Pearson Correlation	-0,055	1	0,033	0,656	0,290	-0,076	-0,192	0,528	0,330	0,152
	Sig. (2-tailed)	0,756	.	0,849	0	0,091	0,666	0,285	0,002	0,070	0,413
	N	34	37	36	35	35	35	33	33	31	31
The Importance of Foreign Tourists for the Hotel	Pearson Correlation	0,349	0,033	1	0,157	0,322	0,324	0,042	0,271	-0,228	0,310
	Sig. (2-tailed)	0,043	0,849	.	0,374	0,060	0,058	0,818	0,127	0,217	0,090
	N	34	36	36	34	35	35	33	33	31	31
The Importance of Domestic Groups of Tourists for the Hotel	Pearson Correlation	0,024	0,656	0,157	1	0,518	0,093	-0,082	0,316	0,311	-0,108
	Sig. (2-tailed)	0,897	0	0,374	.	0,002	0,600	0,649	0,078	0,101	0,577
	N	32	35	34	35	34	34	33	32	29	29

The Importance of Foreign Groups of Tourists for the Hotel	Pearson Correlation	0,180	0,290	0,322	0,518	1	0,119	-0,010	0,073	0,312	-0,005
	Sig. (2-tailed)	0,317	0,091	0,060	0,002	.	0,502	0,956	0,691	0,093	0,981
	N	33	35	35	34	35	34	33	32	30	30
The Importance of Seminars with Less than 100 Participants for the Hotel	Pearson Correlation	0,800	-0,076	0,324	0,093	0,119	1	0,662	-0,382	-0,096	-0,121
	Sig. (2-tailed)	0	0,666	0,058	0,600	0,502	.	0	0,028	0,613	0,524
	N	33	35	35	34	34	35	33	33	30	30
The Importance of Seminars with More than 100 Participants for the Hotel	Pearson Correlation	0,455	-0,192	0,042	-0,082	-0,010	0,662	1	-0,250	-0,206	-0,284
	Sig. (2-tailed)	0,010	0,285	0,818	0,649	0,956	0	.	0,167	0,293	0,142
	N	31	33	33	33	33	33	33	32	28	28
The Importance of Health Resort Guests for the Hotel	Pearson Correlation	-0,476	0,528	0,271	0,316	0,073	-0,382	-0,250	1	0,122	0,332
	Sig. (2-tailed)	0,007	0,002	0,127	0,078	0,691	0,028	0,167	.	0,528	0,079
	N	31	33	33	32	32	33	32	33	29	29
Guest Target Group with the Shortest Stay in the Hotel	Pearson Correlation	-0,259	0,330	-0,228	0,311	0,312	-0,096	-0,206	0,122	1	0,073
	Sig. (2-tailed)	0,175	0,070	0,217	0,101	0,093	0,613	0,293	0,528	.	0,696
	N	29	31	31	29	30	30	28	29	31	31

The Shortest Stay of The Target Group (Number of Days)	Pearson Correlation	-0,007	0,152	0,310	-0,108	-0,005	-0,121	-0,284	0,332	0,073	1
	Sig. (2-tailed)	0,972	0,413	0,090	0,577	0,981	0,524	0,142	0,079	0,696	.
	N	29	31	31	29	30	30	28	29	31	31

*. Correlation is significant at the 0.05 level (2-tailed).
**. Correlation is significant at the 0.01 level (2-tailed).

Source: Ivankovič, 2004, p. LX

Živa Čeh

Non-native Speakers Communicating in English: The Language of Tourism

Abstract

In the tourism industry foreign languages perform a key role. Not only in the communication between a professional and a tourist, but also in the communication between professionals from different countries. It happens very often that the English language is used even when there are no native speakers of English involved in the conversation. In spite of the fact that it is used as the lingua franca, we still want to speak the language similarly to the way native speakers do. However, getting closer to native speaker competence is not an easy task at all. The better we speak a foreign language, the more we notice how difficult it is to get closer to native speakers. One of the problems we have to deal with is the necessity of learning various types of word combinations.

1.   Introduction

In my paper I intend to touch upon some thorny issues confronted by non-native speakers of English. Although MacKenzie (2003:59) says that non-native speakers of English are »guestimated« to outnumber native speakers by about four to one and that an unknown number of these will never need to communicate with native speakers of English, the vast majority of learners in English departments still tend to want to learn the language in such a way that it resembles a native version of English. On our way to getting closer to the language that native speakers use, word combinations play a very important role.

2. Two Models of Interpretation

There are two different ways of explaining how meaning arises from text. The first one is the open choice principle, which is a way of seeing text as the result of very large number of choices. The only restraint is grammaticalness. It is often called a slot-and-filler model; texts are seen as a series of slots, which have to be filled from a lexicon. Gabrovšek (2000:188) speaks about free combinations that are simply sequences of elements joined in accordance with the general rules of syntax that freely allow substitution. For example She/Mary/My aunt does not like/has never liked her parents/mum and dad. However, we would not produce normal text simply by operating by the open-choice principle. That is why the second one, the idiom principle, is put forward. According to this principle language users have available to them a large number of semi-preconstructed phrases that constitute single choices. Word combinations, among them collocations, illustrate the idiom principle.

Cowie (1998:57) explains that phraseology is the study of the nature and distribution of words that are not completely free in combination. There is increasingly strong evidence that phrasal items of various sorts account for the larger proportion of words in much of language production and therefore constitute a significant proportion of a speaker's vocabulary.

Greenbaum (1996:427) points out that in much of our everyday language we draw on prefabricated items rather than selecting words individually.

3. Word Combinations

When we start learning a foreign language we first concentrate on words and some basic structures. In the beginning we are very happy with the progress we make. We learn numerous words and are able to speak about various topics. Nevertheless, to be able to communicate in a foreign language we also have to know how to put words together. After we have acquired the basics of a foreign language we need to become interested in word combinations. There are numerous types of word combinations and fuzzy borderlines between them. No matter how we define different groups of word combinations, varieties of them are used all the time, some of them more and some of them less frequently.

Gabrovšek (2000:188) opts for a basic division into idioms and collocations as contrasted with compounds on the one hand and free combinations on the other.

4.   Idioms

Idioms are presented frequently in course books, although in everyday life they may not be used very often. They are fixed word combinations whose meaning is usually opaque: we normally cannot guess what an idiom means just by understanding the meaning of the constituent words. When we say, for example, that somebody kicked the bucket, we are not saying that they kicked a bucket rather than, say, a ball, but that they died. Similarly, when mentioning that it is raining cats and dogs, we are not implying that cats and dogs are falling from the sky, but simply that it is pouring with rain. Native speakers do not use idioms very often, but when they do, they use them to evaluate, sum up, comment, establish rapport, construct dialogue, be funny and so on. Non-native speakers of a foreign language rarely use them appropriately and many are under the wrong impression that idioms are frequently used. Native speakers of English say that non-native speakers seldom get idioms right. Native speakers of English would find it quaint to comment on the weather by saying it is raining cats and dogs. They would more probably say (Prodromou, 2003:10) it's coming down in stair-rods, it's chucking it down, or it's coming down in buckets. Thus, the idioms non-native speakers remember from course books are sometimes rare and unnatural for native speakers. Although we are told that idioms are relatively fixed expressions, the author also gives example of native speakers bending and breaking the rules of idiomaticity to create special effects. On seeing that it was beginning to drizzle, a native speaker commented it's raining kittens and puppies.

Non-native speakers have different problems with different types of word combinations. Idioms, the word combinations whose meaning is not transparent, may be difficult to translate from a foreign language into one's mother tongue. If we are not aware that a particular word combination is an idiom we can translate it literally word by word. I am sure you will have spotted some very funny translations in film subtitles. Those I have seen include it 'beats me' translated as 'something hit me' instead of 'I don't know' 'and it fell off the back of a lorry' translated as 'it fell off the lorry' rather than 'it was stolen'.

5.   Collocations

Collocations, on the other hand, are used all the time. They are word combinations that are loosely fixed, and whose meaning is transparent. To understand a collocation, it is usually enough to understand the meaning of the words it consists of. In order to become a proficient speaker of a foreign language, one needs to come to grips with collocations. Putting words together can be a very difficult task: often there are no rules about how to do that and no reasonable explanation readily available. We use certain word combinations because native speakers do so. Common examples involve the verbs make and do. Most non-native speakers find it easy to remember that we make a bookcase and do a job, but they may be puzzled by the fact that we also make a phone call, make progress, and do a favor.

The definition of collocations sometimes varies from author to author, but they are usually described as loosely fixed and semantically transparent word combinations. Benson (1997:XV) writes that a grammatical collocation is a phrase consisting of a dominant word (noun, adjective, verb) and a preposition or a grammatical structure such as an infinitive or clause. For example, blockade against, a pleasure to do, an agreement that, by accident, angry at, necessary to do. Grammatical collocations are more deterministic and more often found in dictionaries. Lexical collocations are more problematic for non-native speakers and also more difficult to find in dictionaries. They consist of different combinations of nouns, adjectives, verbs and adverbs. For example, come to an agreement, reject an appeal, strong tea, alarms go off, a swarm of bees, deeply absorbed, affect deeply.

In the introduction to the Oxford Collocations Dictionary (2002, vii), Diana Lea defined collocation as the way words combine in a language to produce natural-sounding speech and writing. Native speakers do not find word combinations difficult, strong wind and heavy rain come naturally to them, while non-native speakers may feel a little bit lost and wonder why it is that we do not say *heavy wind and *strong rain. She sees word combinations ranged on a cline from the fixed and idiomatic, such as not see the forest for the trees, to collocations, such as see danger, see reason, see the point. She adds that the collocationally rich language is also more precise. I would like to emphasize that learning collocations does not mean learning new words. It is about putting together the words we already know. (* indicates wrong usage)

Collocations, word combinations whose meaning is in most cases transparent, should not cause any problems to non-native speakers when translating from English into their mother tongue (decoding). However, the same cannot be said for translating from the non-native speaker's mother tongue into English (encoding). Many words can be found in a dictionary, but how do we put them together and where do we find collocations?

If we use a wrong collocation in English, it does not necessarily cause a communication breakdown. We may get the message across, but native speakers will find the way we put the words together unusual, inappropriate or funny. The correct use of collocations may be a very thin line that non-native speakers have to cross to become fully proficient speakers of English, but needless to say this line is very difficult to cross. For instance, Slovenian speakers of the English language will make the mistake *to be interested for because of the collocation in the Slovene language - they automatically translate the Slovene preposition. In a way, the Slovene collocation misleads them into using the wrong preposition in the English collocation. Many Slovenian speakers of English will for the same reason say *sweet water instead of fresh water, *typical for (typical of), *to cook coffee (to make coffee), *to go on coffee (to have a cup of coffee), *interested for (interested in), *good in (good at), *allergic on (allergic to), *married with (married to), *depend of (depend on), *black bread (brown bread), *black wine (red wine) because they simply and wrongly translate the Slovene expression. Gabrovšek (1998:129) mentions some more mistakes typical of Slovenian speakers of English: *a clock shows time (a clock tells time), *a high age (an advanced age), *angel guardian (guardian angel).

Typical mistakes vary from one group of speakers of a foreign language to another. To avoid these types of mistakes we have to find out in which dictionaries collocations can be found and under which word.

6. Dictionaries

Some linguists say non-native speakers should never translate into a foreign language, as they will never get it exactly right. This might be true in terms of collocations. It is relatively easy to find individual words in a bilingual dictionary, but more difficult to find information about word combinations. Dictionaries usually offer some more grammatical and less lexical collocations. It is very important that a bilingual dictionary for encoding includes collocations that usually cause problems for a particular group of speakers of English. Thus in a Slovene-English dictionary one would expect to find the collocations Slovenian speakers of English often have trouble with. If the collocation we want to look up cannot be found in a bilingual dictionary, we can consult a monolingual English dictionary. Some of them, especially the newer ones, offer plenty of grammatical and lexical collocations. Sometimes it seems that the newer the dictionary is, the more collocations it offers to its users. If we still haven't managed to find the collocation we are looking for, there are monolingual dictionaries of collocations offering plenty of word combinations. It might be a bit difficult for some users to learn how to use a dictionary of collocations; grammatical collocations can be found under a dominant word, lexical collocations under one of the words in a word combination. With some dictionaries there are also supplementary exercises available which teach the users of the dictionary how to use it. The only problem with a monolingual dictionary of collocations is that it is written for all non-native speakers of the language (Slovene, Dutch, French…), but not for any of them in particular. As a result it might happen that a Slovenian speaker of English will find plenty of collocations that do not cause problems for Slovenian speakers of English, but none of those that are problematic for Slovenian speakers of English.

In my research (Čeh, 2001:101), I looked up 100 random choice collocations that are typical of the language of tourism. 66 of them are semantically or structurally different from Slovene translational equivalents and as such problematic for Slovenian speakers of English. Just 12 of them can be found in the Slovene-English dictionary. When searching for the collocations that are problematic for Slovenian speakers of English in an English dictionary of collocations I found out that 12 of them were not included at all. Among them are such expressions as a provisional booking, weather permitting, to allocate a room, a bucket and spade holiday, dry lease, a low-level front desk. It should be mentioned that a dictionary of collocations is not a specialized one and does not necessarily include word combinations from the language of tourism.

Logically, a bilingual dictionary of collocations would be the right reference, but few nations can afford and actually have one. Slovenia does not, and in the short term probably will not get one. For this reason it is even more important to apply the contrastive approach in English lessons.

7.   Lexical Phrases

Lexical phrases are word combinations that can be fixed or variable and perform a very important function in the text. They can be short, a __ ago, or long the __er, with empty spaces that have to be filled in or fixed. They may cause problems to non-native speakers of English, as there are several things to remember. First of all, non-native speakers have to know which function they perform in a sentence and they also have to remember whether they vary or not. And non-native speakers should remember them as a whole, as chunks of language.

I do not intend to deal with lexical phrases in detail in this paper. I only want to illustrate how important they are by mentioning some of them and the functions they perform in a text. How do you do is used when we are being introduced. For the most part is a qualifier, in a nutshell is a summarizer, by the way is a topic shifter, I'll say is an agreement marker, hold your horses (disagreement marker), at any rate (fluency device), what on earth (marker of surprise), so long (parting), for that matter (relator), so to speak (fluency device), beside the point (evaluator), strictly speaking (evaluator), you know (clarifier). Some phrases are not grammatical: as it were (examplifier), so far so good (approval marker), all in all (summarizer), by and large (qualifier), not on your life (disagreement marker), once and for all (summarizer), in part (qualifier), in essence (summarizer), nevertheless (relator). Some of them are longer expressions: a watched pot never boils (advice), the public seldom forgives twice (warning, disapproval marker), to name just two. By taking into consideration their functions we realize how important they are and how fluent we become in a foreign language when we remember them and use them correctly.

It is quite difficult for a non-native speaker to remember which of them are variable: a __ ago: a day ago, a long time ago (temporal relator); to __ this up: to tie this up, to wrap this up (summarizer); as I was __: as I was saying, as I was mentioning (topic shifter); in __: in short, in sum, in summary (summarizer); good __: good morning, good evening (greeting); yours __: yours sincerely, yours truly (closing); __ as well as __: this one as well as that one (relator); as far as I __: as far as I know, as far as I can tell (qualifier); to make a (very) long story (relatively) short (summarizer).

Some phrases used to construct sentences are:

I think that __: I think it's a good idea (assertion).

> Not only __, but also __: not only are most spiders harmless, but they are also beneficial (relators).

> My point is that __: my point is that gravitational force is by far the weakest (summarizer).

> I'm a great believer in __: I'm a great believer in putting money away for a rainy day (evaluator).

> Let me start by/with __: let me start by mentioning a few of the most important things (topic marker).

As I have already mentioned lexical phrases should be remembered as ready-made chunks of language because this is the only way they can be instantly retrieved and used.

8.   Course Books for Students and Teaching

It would not make any sense to start teaching collocations before students have a fairly good knowledge of a foreign language. Secondary schools are the right places to introduce collocations. In many course books for secondary schools there are idioms. We can use them as a springboard for discussion of different types of word combinations and point out the differences between them. Even better, many course books offer exercises using collocations, mostly different word combinations with prepositions, or verbs, nouns and adverbs. Using these, we can introduce our students to the two types of collocations and tell them in which dictionaries they can find information about word combinations.

On the tertiary level we have to pay special attention to ESP (English for Specific Purpose). A large part of special vocabulary is collocational and to be able to speak about our field we have to know the relevant collocations. Students can be encouraged to study collocations using any available materials, course books, articles, videos, etc. After giving some general information about different types of collocations, we can apply the contrastive approach and translation.

Just as English monolingual dictionaries are written for all groups of non-native speakers of English, numerous course books are sold all over the world. What was said about dictionaries goes for course books as well. They cater to all non-native speakers of English but to none in particular. There will be more and more collocations included in course books, but maybe none of those causing problems for our speakers of English. That is why teachers have to collect and add them to their lessons.

In the course book I wrote for students of tourism (Čeh, 2003:22) I included some collocations. At the same time I started collecting the mistakes my students make in their papers, especially those including word combinations, and began to use them in English lessons. Readers of the article who speak Slovene will immediately realize that the mistakes collected here were made under the influence of the Slovene language. As it would not make any sense to include just collocations in the exercises, I collected all types of word combinations. My purpose was to make students aware of the mistakes they make and how to correct them.

Here is an example of the text for students.

> Is there a better way of saying it?
>
> Ljubljana is the main city of Slovenia.
>
> It is a very beautiful town.
>
> Tourists who come to Ljubljana can visit a lot of interesting places.
>
> One of the things you must not miss is grad – the castle which stands on a small hill over the city.
>
> You can also visit some festivals and exhibitions or go for a walk around the castle.
>
> There is a beautiful view on Ljubljana.
>
> You can see the three bridges built by a famous architect Jože Plečnik.
>
> There is also the river Ljubljanica.

The old part of the town is very interesting because of the old houses and narrow stone streets.

There are many bars and shops where you can buy different things.

Some souvenirs are typical for Slovenia and they are very beautiful.

For example lace from Idrija, bee hives with paintings, candles made from bee wax and cookies made with honey.

There are usually some love messages on them.

There you can also buy a lot of things made from wood, like tools for gardening, things for the kitchen and toys for children.

The most interesting are horses for small children.

In the evening you can go to the cinema or visit one of the many performances in the theatres or concert halls.

A visit to Ljubljana will always stay in your memory.

We correct the text together and in this way students become aware of different collocations in a foreign language. We would usually end up with a text very similar to the one below.

Ljubljana is the capital of Slovenia.

Tourists who come to Ljubljana can take an interesting sightseeing tour.

One sight not to be missed is grad – the castle that stands on a small hill above the city.

You can also go to some festivals and exhibitions or go for a walk around the castle.

There is a marvellous view of Ljubljana.

You can see the three bridges designed by the famous architect, Jože Plečnik.

The old part of the town is very interesting because of the old houses and narrow cobbled streets.

There are many cafes and shops where you can buy a wide variety of things.

Some souvenirs are typical of Slovenia and they are very pretty.

For example lace from Idrija, beehive panels with paintings, candles made of bee's wax and cookies made with honey.

There are usually some love poems on them.

You can also buy a lot of things made of wood, like tools for gardening, utensils for the kitchen and toys for children.

The most delightful are rocking horses for small children.

In the evening you can go to the cinema or attend one of the many performances in the theatres or concert halls.

You will always remember your visit to Ljubljana.

Using wrong collocations is not the only problem non-native speakers might have. When we do not know a collocation in a foreign language, we can avoid it by using descriptions. To encourage students of tourism to learn and use English collocations I ask them to do the following exercise.

Is there a better way of saying it?

Cruising was before thought of as a holiday you take just one time in your life.

It now competes with holidays that include more components and when you buy it you really get what you paid for.

You must not forget that it is a holiday that includes all meals.

Cruise companies give passengers food of good quality and you can usually get it at any time.

Another reason for the fact that cruising is becoming more popular is that you can choose between different offers.

There are cruises to each country of the world offering what different age groups want to buy.

More and more tourists want to buy cruises which offer something to people with a particular interest.

They organize lectures on a ship and they include different topics.

You do not have to pay extra to attend lectures.

It is now possible to choose between so many different topics that it is important to choose carefully.

Luxurious boats will take you to many different places tourists go to.

All cabins have their own bathroom and air-condition.

Cruise ships usually offer outside swimming-pools and also some other equipment.

The picture most people have of cruising has changed very much.

We end up with a text similar to the one below.

> Cruising was formerly thought of as a once-in-a-lifetime holiday.
>
> It now competes with package holidays for value for money.
>
> You have to bear in mind that it is a full board holiday.
>
> Cruise companies provide quality food usually available round the clock.
>
> Another reason for the growing popularity of taking cruises is the choice on offer.
>
> There are cruises all over the world catering to all age groups.
>
> There is a growing demand for special interest cruises.
>
> They organize lectures on board that cover a wide range of topics.
>
> Lectures are free of charge.
>
> The choice is now so wide that it is important to select carefully.
>
> Luxurious boats will take you to numerous holiday resorts.
>
> All cabins have en-suite facilities and air conditioning.
>
> Cruise ships usually offer outdoor swimming pools as well as some other facilities.
>
> The cruising picture has changed dramatically.

It has been already mentioned that collocations are word combinations with which native speakers decide what is acceptable. I am not a native speaker of English and it would be wrong to think that I am the one to make that decision. For this reason I asked a native speaker to help me with the exercises I prepared for my students.

Collocations are everywhere, in the texts we read, the tapes we listen to and the videos we watch. We do not need any special materials to find them. After watching a video about national parks, for instance, numerous word combinations can be collected. Two main purposes of national parks, conservation of the landscape, promotion of the area for public enjoyment, the national park was designated in 1929, there are several honey-pot sites, prevent erosion, reinforce the river banks, the clash between local population and visitors, traffic congestions, build footpaths, they are trying to prevent pollution are just some of them. Such expressions are of great value to students as they are ready-made chunks of language and can be used immediately without having to put the words together on our own.

Collocations are also on objects we use in our everyday lives. Taking a tea box into our hands can equip us with the following word combinations: the company was established, tea and coffee merchants, a special blend to awake and invigorate, loose tea, tea from several origins, a bright full-bodied amber tea, brewing instructions, best served with milk, set up a company, trade continuously, a unique record, such teas give unequalled pleasure and satisfaction. If we are teaching the language of tourism in Slovenia we can immediately switch to the advertising slogan of the Slovene Tourist Board, which is 'Slovenia invigorates', or maybe to business English and the expression advertising slogan. If we are speaking about everyday life, loose tea, tea bags, tea cosy and tea caddy can be mentioned, and so on. Again, these word combinations will be ready-made and if they are remembered as such they will not have to be put together word by word.

Farghal and Obiedat (1995:315) start their article by saying that collocations are an important yet neglected topic. Their research shows that learners of English as a foreign language are seriously deficient in collocations and as a result resort to strategies of lexical simplification.

Nesselhauf (2003:223) investigated the use of verb-noun collocations (take a break) by advanced German-speaking learners of English. She identified the type of collocations that learners found most difficult and the role of the learners' mother tongue. In verb-noun collocations the focus of teaching should be on the verb, since it is the verb that causes the greatest difficulties. We should teach collocations that are frequent, that are different from collocations in our mother tongue and that allow for the change of one word in a word combination since these are the collocations that most learners find most difficult.

Shei and Pain (2000:167) agree with Farghal and Obiedat (1995:315) that although collocations are important and difficult they have been largely neglected in foreign language learning. They developed an on-line correcting program used to detect collocational errors and offer standard collocations from a large corpus for reference.

Howarth (1998:24) points out that advanced learners of English need to have command of a wide range of complex lexical units, which native speakers process as prefabricated chunks. They are fixed or semi-fixed expressions. The author adds that there is a lack of detailed description of learners' phraseological performance.

Rather surprisingly, De Cock (2000:51) found out that some learners of English overuse prefabricated sections. It shows that non-native speakers of English rarely get it right when it comes to prefabricated chunks.

Lewis (2000:12) advises teachers of English as a foreign language to react immediately to collocational mistakes such as *strong smoker*. We can immediately come up with correct collocations of the word smoker, for instance heavy/occasional/chain/non-smoker. He also believes that knowing a word means knowing its collocations and illustrates it with the words wound and injury. It is most useful for learners of English to know that it is correct to say stab wound and internal injuries, but not *stab injury* and *internal wounds*. Similarly in the language of tourism the difference between the words destination and resort does not lie in their dictionary definitions - destination of someone or something is the place to which they are going or being sent, a resort is a place that a lot of people go to for a holiday - but in their collocational fields. We can say: favourite/popular/holiday/tourist destination, and also: favourite/popular/holiday/tourist resort. But it does not mean that, if we can say health/purpose-built/ski/spa/mountain/seaside/hotel resort, that we can also say *health/purpose-built/ski/spa/mountain/seaside/hotel destination*.

9.   Conclusion

To sum up, no written or spoken English text is without word combinations. They are numerous and used all the time. Choosing the right one will make the non-native speaker's language sound more natural, fluent and precise.

References

Benson, M. et al. (1997): The BBI Dictionary of English Word Combinations. Rev. ed. Amsterdam: John Benjamin's Publishing:XV.

Cowie, A. (ed.) (1998): Phraseology: Theory, Analysis and Applications. Oxford: OUP:57.

Čeh, Ž. (2001): Angleško-slovenska protistavna leksikalna analiza jezika turizma : Kolokacije v teoriji in praksi. Magistrsko delo. Ljubljana: Univerza v Ljubljani, Filozofska fakulteta:101.

Čeh, Ž. (2003): Course Book for First Year Students. Portorož: Turistica, Visoka šola za turizem:22.

De Cock, S. (2000): Repetitive Phrasal Chunkiness and Advanced EFL Speech and Writing. In: Mair C. and Hundt M. (eds.) Corpus Linguistics and Linguistic Theory, Amsterdam: Rodopi:51-68.

Farghal, M., Hussein, O. (1995): Collocations: a Neglected Variable in EFL. IRAL 33, (4):315-331.

Gabrovšek, D. (1998): Coping with Stubborn Stains and Persistent Headaches – for what it's Worth: Word Combinability in Action. Vestnik 32, (1-2):111- 154.

Gabrovšek, D. (2000): Phraseology Galore: Words and Their Combinations. Vestnik 34, (1-2):183-238.

Greenbaum, S. (1996): The Oxford English Grammar. Oxford: Oxford University Press:427.

Howarth, P. (1998): Phraseology and Second Language Proficiency. Applied Linguistics 19, (1):24-44.

Lewis, M. (ed.) (2000): Teaching Collocation. London: LTP.

MacKenzie, I. (2003): English as a Lingua Franca and European Universities. The European English Messenger 12, no 1:59-62.

Nesselhauf, N. (2003): The Use of Collocations by Advanced Learners of English and Some Implications for Teaching. Applied Linguistics 24, (2):223-242.

Oxford Collocations Dictionary (2002): Oxford: Oxford University Press:vii

Prodromou, L. (2003): Idiomaticity. English Teaching Professional no. 27:10-12.

Shei, C., Pain, H. (2000): An ESL Writer's Collocational Aid. Computer Assisted Language Learning 13, (2):167-182.

Goran Vukovič, Jure Meglič, Brane Šmitek

Model of an Interactive Presentation for a Tourist Destination Offer

Abstract

The article presents a model for the multimedia promotion on the World Wide Web of regional tourism facilities. The model is designed on the basis of research conducted by the Slovenian tourist organization, Tourist association of Slovenia and on the strategic guidelines for developing Slovenian tourism. The concept of up-to-date tourism promotion necessitates presentation serviceability, the possibility of on-line reservations and do-it yourself individualized tourism destination offers. The presented model also offers tourism companies and associations the advantage of promptly accompanying and analyzing new tourists demands and customizing the tourist products within a very short response times.

Key words: tourist destination, tourist management, tourist product, multimedia World Wide Web portal

1.   Introduction

The tourist industry is one of the greatest acquisitions for an economy in the 20th century. As a result of its complexity it is an exceptionally demanding part of the service industry because it intertwines different areas of human activity. As the contractor and tourist, man is still the decisive factor in determining the success of a tourist offer, no matter how high the standard is. Constantly changing local and international conditions as well as the faster cycle of development in the tourist industry have also forced tourist destinations to think and operate in a new way. The current traditional structures have become more and more inadequate in an increasingly competitive global environment and cannot adequately compete for the inclination and attention of the consumer.

In terms of tourism, the Gorenjska region represents the third most important region in Slovenia and is divided up into two destinations: The Julian Alps (the northern and western part of Gorenjska) and Carniola (the southern and eastern part of Gorenjska). Undoubtedly, the tourist industry is a significant economic and social factor, as shown by the following data from 2003:

Table 1: An overview of the key economic indicators in the tourist industry (Meglič, 2004:221)

Indicator	Gorenjska	Slovenia	EU (15)
Territory	2,137 km2	20,273 km2	3,9 mil km2
Population	0.2 mil	2 mil	454 mil
GDP/person	10,338 €	11,775 €	21,197 €
% GDP in the tourist industry	20.72%	9.11%	11.00%
% employed in the tourist industry	11.81%	5.86%	9.00%

Slovenian tourism has one of the greatest output branch multiplicators (1.8) as it: (1) connects many other industries in the economy (2) accelerates regional development and (3) increases the economic value of the area from the standpoint of appreciating nature and cultural heritage.

Modern technologies have been surpassing boundaries up to the current times and opening up completely new methods of presentation and other marketing activities. New consumer and market profiles have been developing, which also demands adequate adaptability from tourist service providers. The modern tourist is oriented towards spending shorter and active holidays throughout the year, values an undamaged environment and services that are related to wellness alongside a high level of individualism. This is why service providers do not identify the needs and desires of modern tourists fast enough even though modern technology, especially interactive portals on the World Wide Web, enables them to do so. The presented model of interactive tourist destination offers also enables the dynamic shaping of individualized offers to potential tourists whilst at the same time providing online information on the trends and needs of the modern tourist to service providers of tourism.

2.   Shaping Tourist Destination Offers

One of the principal aims in the tourist industry is to develop tourist products with added value for current and potential target markets in such a way that tourist destinations and their communities can obtain social and economic benefit (Yoon, 2002:43). Managing tourist destinations demands a certain way of connecting tourism, the public and the social sector in the form of a public-private partnership. The government of the Republic of Slovenia wanted to achieve this in 1998 with its law on accelerating tourism. However, the current conditions in the organization of tourism at local and regional levels show that over the last three years the law has not been successful at essentially connecting and encouraging cooperation among subjects in tourism.

Local tourist organizations in developed tourist areas, which were established before the law was passed, were the most successful in developing destination management. They were successful in connecting at least part of the tourist subjects in local and regional tourist areas. However, also in these cases, local tourist organizations were mostly successful in developing only promotional activities and did not succeed in shaping other marketing functions in this short period of time, which demands a modern approach to management at tourist destinations. The biggest differences occurred among local tourist organizations, which had started developing before the law on accelerating tourism was passed and those which were established after the law was passed. After 1998, local tourist organizations often emerged in less developed tourist areas and most of them were limited to one municipality, which shows a certain lack of preparedness for interconnection across municipal borders.

Limiting local tourist organizations within the framework of individual municipalities, a lack of development in tourism in these areas and weak preparation of these tourist subjects for partnership cooperation and co-financing their activities were the fundamental reasons for not understanding the real role of local tourist organizations (LTO). The fundamental supposition of the law in 1998 was that the reorganization of Slovenian tourism was the fundamental factor in tourism development. The law defined that the LTO shapes the representative, information and promotional function at local level. The first problem with this approach was developing the status of the LTO's, which in most cases had become public institutions and not institutions for marketing in business, which would have followed the modern principles of setting up businesses, privatization and the socialization of the public sector in tourist areas.

The majority of the 32 local tourist organizations developed into municipal administrative institutions. Local tourist organizations in less developed tourist areas were often not able to ensure a sufficient amount of funds for normal operations. That is why their management used a legal provision on obligatory membership fees but at the same time did not take into consideration the interests and wishes of other local tourist subjects. In some cases they merely took over the activities of the already existing tourist organizations. The marketing department of the LTO often presented a disloyal competitor to private tourist subjects, which caused even greater opposition by its members to cooperating with the LTO.

Explicit public organizational interest in Slovenian tourism at local level caused the public sector to gradually push out the private sector in the area of the overall marketing of the integral tourist products of tourist destinations. The tourist areas had distanced themselves even more from the much desired and legally defined development guidelines in public-private partnership and management at local and regional levels. Although many LTO's achieved significant results in the areas of promotion and connecting local tourist offers, the overall system of organizing Slovenian tourism failed:

- because it wanted to accelerate the tourist offer with organizational changes instead of stimulating tourism programs,
- because it was based on budget funds or obligatory membership fees instead of the market and interest membership,
- because it did not use comparable modern European methods of new public-private partnerships.

2.1 Key Sources in Tourist Destination Offers

The economic success of tourist destinations depends on many factors, which impact the appeal, quality and profitability of a tourist offer. Below is an overview of five of the most significant sources for developing tourist destination offers.

Table 2: Overview of key sources in tourist destination offers (Jonker, 2004:115)

Category of Source	Main Characteristics	Key Indicators
Physical Means	Physical means (sources) of destinations including their natural, cultural and artistic attractions are the main determinants of tourism potential. Also the tourism infrastructure (i.e. roads, communication...) are really important for tourism development. Tourists visit destinations because of special experiences and their expectations are mostly developed as a result of the offered attractions.	Topographical and landscape diversity (i.e., opportunities for swimming, natural reserves, forests, flora and fauna). Nature and the extent of special attractions. Nature and the extent of special positions i.e. world cultural heritage or attractions that have received special recognition. Climate conditions (temperature, rainfall, winds...) Diversity and quality of activities (i.e. events, festivals, shopping, entertainment...) Distance from markets. The quality of the public infrastructure (airports, roads, public transport, communications) Number, quality and diversity of hospitality and tourist offers.
Financial Sources	Financial sources, which are carefully allocated for the development and marketing of destinations as well as a cost factor in the accessibility of a destination.	The budget for the marketing activities of the destination Public funds carefully allocated for projects in the area of tourism Transportation costs, appointment etc. when visiting the destination.
Techno-logical Sources	The capability of destinations to attract foreigners depends on different types of technology (aviation, communication, etc.).	The speed and capacity of telecommunication infrastructures E-business networks. The system of airline routes (connecting flights) The number and quality of reservation-marketable systems
Reputation and Culture	The reputation of a destination's trademark and its degree of recognition in potential markets.	Recognition and value of a trademark on the market The uniqueness of the attractions, events and heritage for which the destination is recognizable. Percentage of returning guests. Seasonal trends. Percentage of guests in the region who are

		visiting the destination.
		Guests' evaluation of the quality of the tourist offer.
		Information circulated by the media (newspaper reports, tourist publications, radio and TV broadcasts.
Human Resources	The competency of those employed in tourism. Commitment, pride in the destination and loyalty among the employees who are in contact with the tourists. Hospitality, friendliness and being polite to foreigners by local inhabitants.	Educational, technical and vocational qualifications of those employed in tourism. Employee wages in tourism. Indicators of stability in the tourism labor force. The quality of guests and ranking of satisfaction Alienation and acceptability of benefits and commitment in tourism among the population The number and frequency of events that endanger the safety of guests.

2.2 Value Chain of a Tourist Destination Offer

A tourist offer is by nature a mix of different activities, whose completeness we must ensure in order to achieve the desired goals. In tourism, success in management is also measured according to the success of achieving the set goals. Well set goals have to be measurable, achievable or stimulating and in the case of negative changes or sudden turns it is also necessary to have alternate goals on the market, which enable re-direction. The following have an influence on achieving the set goals:

- Activities (directives and programs),
- Well-regulated (decision-making, management, communicating, planning, control),
- Sources that are available for achieving the goals, (material and non-material).

The company must define a strategy how to achieve the goals. Alongside the strategy we must develop a plan for performing those activities, which will contribute to achieving the set goals. Therefore, in developing a tourist destination offer it is necessary - besides the key sources - to also include the activities in the so-called valuation chain. From the standpoint of the tourist or the guest we divide the activities into two basic groups: primary and support activities. We further divide them into:

- Primary activities
  - Destination and developing tourist products;
  - Shaping the journeys and itineraries; creation of promotional material; uniting and collective development of city attractions, areas and regions; negotiations and price contracts with suppliers and sub-contractors; the entire development of tourist offers and products;
  - Promotion;
  - Consumer advertising, public relations and promotion; fairs, workshops and visits from sellers, picturesque presentation on the markets; trips and visits by celebrities; media, associations and education;
  - Sales and Distribution;
  - Inquiry-based and information-based electronic messages; an exhibition with presentation material; emphasize special demands; tourist information centers; reservations; payment and the sale of tickets; insurance;
  - Incoming and outgoing logistical activities;
  - Commission for visas and passports; airlines and airline services; returning VAT; emigration procedures; declaration and embarking procedures; dealing with luggage; service during the flight; seat prices and schedules;
  - Airline transfers; taxis; public transport; visiting centers; appointments; catering; traveling; attractions; cars and car rentals; entertainment; health and beauty; sport and recreation;
  - After-sale services;
  - Databases; keeping track of and recording feedback from guests; feedback in different areas and tracking the best.

- Support activities
  - destination planning and infrastructure;
  - public transport systems, roads, airports, railway station, ports, etc.; the size of the telecommunications infrastructure, water, electricity, recreation...; planning destinations, shaping destinations, appearance and utilization of premises; aesthetics, the environment and the quality of progress in society; security and security policies; road signs, information boards, other navigational means; public-private partnership, strategic links, mergers and takeovers; institutional coordination, business deregulation;

o   the development of human resources;

o   the acceptability and alienation of tourism in society; training and education; human resource management, motivation, stimulation, etc.; taking care of guests, hospitality; career plans, personnel development, stability of personnel; attitude towards the labor force and negotiations;

o   the development of products;

o   new airlines and destinations; the development of environmental and cultural sources; upgrading and developing new services and benefits for guests; taking advantage of new markets; new paths, themes, centers, peaks and travel plans; management standards for quality and insurance systems; improving distribution services and managing guests;

o   technologies and systems;

o   information systems for reservations, market research and understanding; managerial systems and instructions; managing water, electrical and other resources; security systems; information systems and communication;

o   linked areas and their maintenance;

o   ensuring support products; fuel, food and drink; contractual services; professional services; other services; real estate and other objects.

2.3   Trends and Orientations

The tourist industry is exposed to strong and growing competition. It is necessary to constantly observe changes in the market carefully and recognize the most important trends and orientations in order to achieve a comparative advantage and successful positioning in the market. The future in the tourist and leisure industries lies in the management of change. Things are no longer static, distribution channels, price strictures, ways of communicating, competition, the social/cultural climate and consumer needs are all changing. Although people are clearly traveling, the ways and reasons are changing and will continue to change in the future. Below we have summarized the trends, which will have the largest impact on the development of tourist products according to research conducted into the strategy of marketing Slovenian tourism (Sibila-Lebe, 2003:13).

2.3.1   Social Trends

Social trends driven by technological and economic improvements within globalization are changing the classical demands of a tourism offer. The most important social trends are:

Table 3: Overview of new social trends in the tourism industry

A Society of Quality Living	Europe is on the path towards being a society of "Quality Living" in which the most significant foundation is health in the holistic sense (physical, emotional and social) and ensuring the preserving ability to enjoy life.
Demographic Changes	Society in Western Europe is experiencing drastic structural changes: households are becoming smaller; the population pyramid is evolving into a mushroom shape. Older people will increasingly become the significant target group; they have more leisure time, the financial means, and are critical and active consumers. Together with the changing "planning life models" the traditional structure of the family is also changing: the number of one parent and patchwork families where parents live together with their own and their children from past relationships, has strongly increased in the last decade and the number of single households has increased up until now to unknown dimensions. People are becoming more and more demanding. They travel in small groups. The average educational level increases the intensity of the travels. They frequently travel on new types of holidays (many short breaks instead of one long one).
Consumer Orientation	In the western world, where the basic needs (food, accommodation and clothes) are satisfied to a large degree, consumer priorities have undergone changes; leisure time activities and entertainment have become increasingly more important.
Price Sensitivity	Most of society is noticeably strongly price sensitive, which is a direct result of the greater and easier access to information. This increasing sensitivity will impact tourism business.
Leisure time	Leisure time in western society has increased over the last few years as a result of shorter weekly and monthly work days. This development can be expected to continue in the future, albeit at a slower rate.
Mobility	The primary reason for positive development in the tourism and leisure markets was the increase in mobility. An increase in mobility can also be expected in the future. As a result of a continual decrease in transportation costs (e.g. flying) trips to far-away destinations will increase. The development of information and communication technology has an influence on obtaining information and methods of reservations.
Individuality	Mass standardization is moving towards individuality even in the area of tourism.

Social Contacts	Industrialization and urbanization go hand in hand with increased isolation and seclusion. It will therefore be necessary to include social contacts and opportunities for individual fulfillment in leisure offers (events, gastronomy etc.)
Ecological Awareness	There has been an increase in ecological awareness and the importance of preserving the natural environment over the last decade. According to the B.A.T – the institute for leisure time (1999) - holidays in the future will often be seen as an event happening in the natural environment: "naturally clean" holidays in an untouched natural environment (27%), "wellness oasis" (20%), "synthetic holiday world" (11%), "exotic" (16%). Slovenia has a good opportunity to become or remain a sanctuary for environmental conservation and ecological tourism.
Political Factor	Integration into the EU and the creation of an internal EU market has facilitated traveling within Europe. With the upcoming expansion of the EU new destinations will appear and new markets will develop.

2.3.2 Leisure Time and Tourism

In terms of changing social trends, the most important factor defining tomorrow's tourism will be the time available to tourists during the year. New time demands are impacted by:

Table 4: Overview of leisure time and tourism

Global tourism market and the leisure market	The tourism industry has substantially grown in the last few years – it has become the global and leading sector in the world economy. Further development of this economic branch is also forecast for the future.
Intensity of Travel and Passengers	The intensity of travel (and the percentage of people within the whole population who travel) has constantly increased in the past.
Traveling Behavior	General social conditions have created a new type of traveling, which can be described as follows:
	The world as a destination: increasing travel to distant overseas destinations;
	Short journeys: more short journeys instead of one long holiday. In many cases the motives for shorter journeys are personal interests (sport, culture, etc.);
	Sharp competition "the seaside vs. the mountains": sharp competition between the seaside and the mountains has been going on for many years. Many mountain villages have developed into real wellness oases, offering sport and entertainment centers for the purpose of following these trends etc.

Predominating Atmospheric Factors	For Germans (and other Europeans) atmospheric factors, such as cleanliness (58 %), comfort (57 %) and friendliness (52 %), rank higher than material goods, which can be bought. Good opportunities for shopping (29 %), opportunities to play sports (27 %) or diversified entertainment (28 %) are only partially interesting.
A Society of New Experiences	With the changes in social behavior and consumer habits the desire for a more intense lifestyle and new experiences is becoming more important. Today, people are more than ever looking for fascination, happiness and positive vibrations (which are usually found during certain events and at different interesting places.
The Development of Demand for Sporting Activities	A strong demand for cycling, hiking, water sports, horseback riding and a continuing strong demand for golf (big golf courses, golf hotels).

2.3.3 Special Trends in Tourism

In terms of a globalized market and individualized consumer demand, the tourists in Gorenjska are looking for the following specialized trends:

Table 5: Overview of special trends emerging in tourism

Bathing – Swimming	• "Bathing" holidays, especially at the seaside are the most popular among Austrians and Germans; • These holidays are linked to the desire for rest, relaxation, laying in the sun, lazing around, etc.; • Bathing and swimming in lakes and/or man-made swimming pools present equally attractive and popular sports activities; • Spa complexes have tremendous tourist significance – they are able to create demand or represent the basic motive for holidays and trips (one day trips)
Hiking	• Hiking still ranks among the most popular sports and leisure time activities for Austrians, Germans and Italians; • Hiking meets popular requirements, such as feeling good about oneself, rejuvenating, experiencing nature, enjoyment and relaxation; hiking combined with trekking and activities in nature is becoming a new sporting trend. • Hiking offers are very diversified – from a simple walk to a fast walk with sticks and Alpine experiences.

Cycling	• Cycling as a leisure activity is gaining in popularity and has already become more important that other sports activities, such as hiking bathing and swimming; • Many cyclist profiles reveal a competitive or recreational cyclist who cycle in order to relax and spend their leisure time performing a sporting activity, which is not too tiring; • Recreational cyclists and trekking cyclists represent the biggest segment with 70% while the more competitively oriented cyclists (e.g. mountain biking) account for the other 30%; The figure of 70% shows • A high level of importance of cycling as a family activity, which enables group experiences (cycling for enjoyment); • Constant expansion of the age range – from 5 to 70 years; • Increasing demand for cycling programs (wine roads, roads to castles etc.); • The daily maximum tour covers 25 to 50 kilometers; • The cycling paths must be separate from the main roads and must have a really good infrastructure; the topography must not be too demanding (no steep hills).
Wellness	• The level of growth from 150 do 200 % makes this segment one of the most dynamic; • Besides the strong increase in demand, the number of companies offering these services has also grown, which increases competition; • Wellness offers contain really diversified programs; clear positioning and standards of high quality for the necessary needs.
Horseback Riding, Golf, Water Sports	• All these activities are witnessing positive levels of growth; • These offers are ideal for niche products and can significantly contribute to the clear positioning of a company, town or small region.

Skiing	• The demand for Alpine skiing has achieved a certain level of maturity – new techniques, such as carving, boarding have stopped the noticeable downwards trend;
	• Competition among ski resorts has sharpened tremendously in the last few years – small resorts have stayed in the background;
	• The main elements for future sport oriented resorts:
	• Absolute assurance of snow (height above sea level!);
	• The attractiveness of the ski center (quality of the lifts, the surface of the slopes, diversity of additional offers);
	• Big skiing regions or skiing regions that are linked;
	• Regions or villages with a limited low-quality offer have less opportunity for success in the market;
	• Other offers alongside skiing (other winter sports, indoor activities);
	• Entertainment infrastructure

2.4   Tourist Products

Nature and the meaning of the word tourism are often abused. In literature many explanations of the words appear or word associations such as "tourism", "tourism industry", "tourist product" and "tourist destination". Today, as a result of its economic significance, many authors use tourism or tourism industry in the widest sense to mean all activities linked with tourism (Baretje&Defert, 1972:66; Leiper, 1979:86; Smith, 1988:33). A tourist product is a conglomerate of different activities, services and benefits, which form the entire tourist experience. This conglomerate consists of five components: destination attractions, destination benefits, accessibility, image and prices. Tourist destinations can be defined as areas with different natural or artificial particularities, which attract foreigners and tourists (Georgulas in Jenkins&Tosun, 1996:196).

With its biotic diversity and abundance of nature and cultural heritage, Slovenian has exceptional opportunities for developing typical, marketable, interesting and integrated, high-quality tourist products that could set it apart from other competing tourism countries. The main reasons for Slovenia's lagging behind in this area are its lack of business innovation and adaptability, poor marketing qualifications and a low level of education of the tourist workforce. Some of the fundamental weaknesses of Slovenian tourism, in terms of guest expectations, are the lack of attractive tourist products and the lack of service quality.

That is why increasing the competitiveness of Slovenian tourism is especially dependent on the development of new interesting products and increasing the quality of the tourist offer. Fundamental strategies aimed at developing new products and increasing the quality of tourist services and integrated tourist products include:

- Stimulating the development of innovative, integrated tourist products and additional tourist services, which contribute to the increase in the quality of integrated tourist products in tourist areas,

- Implementing standards for measuring the quality of business excellence and their harmonization with the rest of the European tourist countries,

- Stimulating the implementation of other internationally acknowledged standards of quality in tourism.

As a result of the richness of its natural environment, the Gorenjska region has good conditions for satisfying the main motives of travelers from the principal originating markets. Swimming, bathing, hiking, cycling, health and wellness are offers, which must be the main promotional themes in the future, as considerable marketing potential still exists in this area. Besides this there are motives, such as business tourism (MICE segment) and casino-entertainment tourism, which must be addressed. As a result of the current structure of ski tourism it does not represent any potential in the target markets outside of Slovenia.

3. The Use of Information Communication Technology for Planning Destinations

Modern tourists usually still avail themselves of travel agency services when first visiting a certain destination. The quality of the program equipment is also exceptionally important in e-tourism (Leskovar, 2000:495). Destinations that occur to them are usually of a more personal nature. Groups are smaller with similar interests and needs. Their visits are planned for a longer period of time and usually avoid tourist destinations that are very popular or really crowded. This is why they plan the trip themselves or together with other participants wanting to visit the destination. Of course, their planning necessitates as much information about the destination and the opportunity to use e-services for logistical support of the visit. This is where modern information communication technology plays a significant role.

3.1   E-Tourism and the Role of ICT

The Internet and the development of e-business have also had a tremendous influence on tourism. Modern information communication technology gives the user direct access to tourist service offers. As a result, this lowers the operating costs. In general, the leisure industry uses the opportunities offered by modern information technology to a large extent. Tourism is becoming the leading user of applications linked with e-marketing, e-sales and other e-services, which are needed for effective business operations. In Europe, the usage of modern information communication technology in tourism by both older members and newly accepted members of the EU is remarkable.

E-business research has shown that four areas of information communication technology are significant in the tourism industry (e-Business W@tch, 2004):

- e-marketing, which uses world wide web technology for advertising tourist destinations on the web or with the aid of e-mail;

- cooperation between agencies and end-users when planning online products;

- e-business increases the extent of sales over the Internet, which enables the availability of such services directly to the selected groups of buyers;

- e-management with customers enables tourist organizations to gain regular customers with the successful sale of tourist products and, in doing so, to increase sales and profit.

The level of usage of basic information communication technology, such as computers, access to the Internet, e-mail and the World Wide Web, has progressed tremendously (e-Business W@tch, 2004) since 2002; however, the tourism industry still lags behind other branches of industry in terms of information communication technology usage. Usage in the latter has expanded mostly in the area of using more specific technology within information communication technology, such as intranet, LAN, WAN and the use of remote computer access. The biggest tourist agencies have made the most progress in the information process despite the importance of information communication technology in the tourism industry and are the only ones that can compare with companies in other branches of industry. According to research (e-Business W@tch 2004), the reason why medium-sized and smaller tourist agencies invest so little is the fairly high cost associated with investing in adequate information technology solutions and, especially, the costs involved in training the company's employees.

In spite of all of the investments made in information communication technology, companies have not made any essential progress in the area of integrating business processes and ERP systems for supporting business operations in tourism. In the near future, we can expect tourist companies to start linking up with others, especially in the purchasing process in relation to their suppliers, as all the companies practically use the same products, which are offered to their guests. Merging and uniformity will demand standardization as this will be the only way to shorten the response time between the points of sale and offer. On the other hand, tourist agencies are taking most advantage of the marketing opportunities (e-Business W@tch, 2004); the best European agencies are already reaching the limits currently offered by the most modern information communication technology. From this standpoint small tourist subjects are still merging more and more in all three pillars of the tourist offer (economic – civil society – the public sector). At the same time, by using systems such as CRM, which enables a high level of individual access by the customer, companies are skillfully individualizing their basic offers and, in doing so, they are raising the tourist product to a higher market level, as shown by the requirement trends of tomorrow's tourists. Companies in France, Italy and Slovenia (e-Business W@tch, 2004) are ranked the lowest among all EU countries in terms of information communication technology usage.

3.2   Specifications Regarding the Size of the Agencies

In Slovenia and in the region of Gorenjska the sector of micro and small tourist agencies is still under development and does not have the same influence on the usage of information communication technology as countries in the western part of the EU since it dominates the mentioned segment of tourist agencies. The implementation of information systems and allocation of funds for e-business in small companies is expensive and, above all, difficult if the right competencies are not in the right place. Big agencies dominate in their usage of information communication technology in areas such as: e-purchasing, functions throughout the whole organization, the integration of information systems and a network infrastructure. Since the majority of information systems for supporting business operations are intended for bigger companies, their usage in micro and small companies is too expensive and their functionality is inappropriate.

With the help of tenders issued by the government of Slovenia, tourist agencies both in the Gorenjska region and in the rest of Slovenia have started to link up and create clusters, which have not yet had any real economic impact, although this can be expected in the near future. In order for micro, small and medium-sized companies to successfully merge it is necessary to simultaneously monitor the usage of technological standards and levels in the implementation of information communication technology, which are suitable for all the partners that are connected. This will definitely be one of the most significant accelerators for expanding business opportunities for the mentioned companies and connections. The research conducted by e-Business W@tch (2004) has shown which areas within information communication technology will be most important for companies in the future. The following table outlines the findings of this research:

Table 6: An overview of the importance of applications for e-business in tourism (e-Business W@tch-a, 2004)

Area of e-business	Importance	Example
Enabling and accelerating remote processes	⊕⊕	Big companies are better at exploiting the advantages offered by remote access in comparison with other sectors than micro, small, and medium companies.
Improving management knowledge with the use of special program solutions	⊕⊗	Using special program solutions for managing knowledge is limited because organizations are not aware of the pure effects. It is presumed that managing knowledge has the greatest potential in the economy.
Automation of internal business processes	⊕⊕	The improvement of business processes in organizations in the tourism sector is of relatively minor significance.
Improving ERP – ERP interconnectedness	⊕⊕⊗	ERP applications should execute the processes in the supply chain more effectively and, as a result, enable shorter response times.
The integration of processes in the supply chain	⊕⊕	The tourism sector has started to focus on managing supply chains, although it has not yet achieved the appropriate results, compared with other sectors.
Decreasing direct purchasing costs via e-purchasing.	⊕⊕⊗	Bigger companies are forcing suppliers towards the stipulated goal, just as in other sectors. In general, the emphasis is on improving the effectiveness of the processes.
The promotion of e-marketing for fundamental services on the world wide web and customers who are connected with the service	⊕⊕⊕⊕	The tourism sector is strongly dependent on the Internet for distributing information, sales and services over the Internet. This is the key to the tourism sector's usage of information communication technology.
The use of CRM applications	⊕⊕⊕⊗	The key area for the tourism sector. Managing personal relations and the individualization of services are key factors.
Increasing the	⊕⊕⊕⊕	Sales over company web pages are the highest

extent of e-business / marketing over the Internet		priority and one of the key activities for organizations in the tourism sector. All companies, regardless of their size, market and constantly upgrade their offers on the web.
Developing a B2B market through the Internet	⊕⊗	Only a limited number of companies in the tourism sector use the market, although this area is expected to become more significant in the future.
The use of e-business standards for exchanging structured data	⊕⊗	A below-average usage for exchanging documents. An element that is not receiving any special attention in the tourism sector at the moment.
Investing in web services and XML standards	⊗	Micro, small and medium-sized companies are refraining from investing in XML applications because they are expensive and comprehending the technology is too slow.
Designing the expanded companies: participating on the web for designing e-products	⊕⊕⊕⊗	More and more organizations are designing partnership links with external contractors in their supply chain in order to design new products, forecast needs, manage the capacity and market the services.
⊕ = minor importance ⊕⊕ = average importance; ⊕⊕⊕ = major importance; ⊕⊕⊕⊕ = very major importance; ⊗ = mixed results, depends on the sub-sector interlinking tourism		

3.3 E-Tourism and the System of Planning Tourist Destinations

Organized complexity is a filter through which we observe and take on the world (Ovsenik&Ambrož, 2000:146). The system of planning destinations and e-tourism engulf a whole series of participants whose demands significantly influence its planning and implementation: Who are the key groups and what are their demands?

3.3.1 Tourists

Tourists on the destination web pages are looking for detailed and up-to-date information about the destination, opportunities for accommodation, interesting things, arrivals and departures, special discounts, experiences of others, etc. Their key demands are:

- fast and cheap access to exact information about the destination with the opportunity to surf further, finding accommodation, accessibility and competitive prices;

- a clear price list, e.g.: price of a room, price for each traveler, etc.;

- quality assurance of the tourist product supported by an evaluation by a certified tourist organization;

- a description of the alternative options for reserving accommodation and the rest of the services that await them;

- the opportunity to use additional services (TV, personal computer, Internet cafe, kiosk, billed telephoning, mobile phone communication, etc.);

- the opportunity to customize the tourist destination offer to their personal needs (including the chance to prepare travel plans for shorter or longer trips),

3.3.2   Trip Organizers

Trip organizers also need more and more information about individual destinations and their needs are very similar to those of tourists. The difference is only that they are specialized and sometimes might have access to additional information that is not available to tourists. Their key demands are:

- an available system for reliable and efficient surfing of vacant capacities and additional services for a larger group of people;

- the opportunity to negotiate prices, dates and discounts etc.;

- the opportunity to quickly contact contractors of tourist services at the destination for a good price;

- ensuring the quality in terms of the standards stipulated by the evaluating organizations;

- simple access to local tourist agents;

- the opportunity to design new local tourist products;

- easy access to important tourist promotional material.

3.3.3   Preparation of Tourist Material

Media companies specializing in presenting tourist destinations also have specific demands, which can be summarized as follows:

- reliable information about newly formed tourist destinations;
- reliable information about the quality of implementing tourist services;
- information about research linked with tourist destinations;
- easy access to experts who know about the conditions in the newly formed destinations;
- easy access to pictures and video material linked with the destinations.

3.3.4   Congress Tourists

Congress tourism has specific demands, which we must take into consideration when planning a tourist destination. We must bear in mind their specific demands because such tourists are specialized and know how to use information communication technology really well. Their requirements include:

- efficient means to search for detailed information on the relevant destination;
- a functioning mechanism to communicate quickly and cheaply with the contractors of tourist services;
- access to promotional material that is needed to organize the congress;
- access to specific information about technological and organizational support for the realization of the congress.

3.3.5   The Tourist Industry

The fundamental business demands are fairly simple – more work, quality services, low operational costs and greater profit. These simple demands are backed up by specific requirements such as:

- a good price for access to an efficient distribution system;
- the opportunity to control distribution channels;
- the opportunity to control prices and the availability of tourist products;
- the opportunity to control automatic reservation verification;
- an offer to an intermediary with the opportunity to negotiate the price and realization of the service;

- a safe and cheap system for payment of the services;

- a clear and modern way of using additional services, which are needed;

- e-business support;

- the opportunity to link up and cooperate with the rest of the participants in the tourist industry;

- taking into consideration and analyzing tourist needs as a basis for designing new tourist offers.

3.4   The Opportunities for Using E-Promotion in Tourism

As communication networks, which are developing faster than ever, improve further, new opportunities for using information communication technology are emerging. Individual and single organization information systems are becoming a thing of the past. Today, interconnected information systems are crucial to enable users to search for up-to-date, quality and useful information. This holds especially true for a tourist product, which has to be complex and must offer the user all the services they need upon their arrival, stay and departure at the tourist destination they have chosen to match their parameters. Connecting all the tourism factors in one area is a necessity, and must encompass the region, state or an even wider area. As a result of the increase in mobility, modern tourists have become more demanding and no longer wish to spend their holiday in just one place but want to experience a tourist destination in its entirety.

Uniting all these demands - with the primary aim of designing an e-tourism system by interlinking the individual factors - produces the following model.

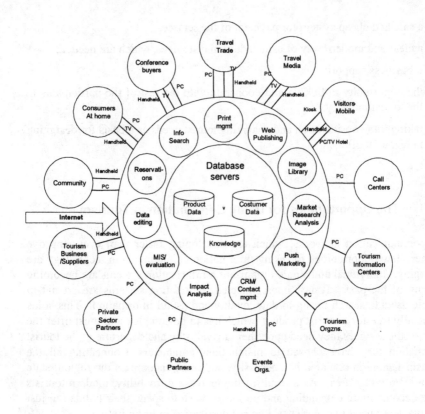

Figure 1: Destination e-business system model [based on a version prepared by TEAM for Western Cape] (Britten, 2002:55)

4.   Conclusion

To successfully establish and design a destination with its own identity despite the many different offers in the same geographic location it is necessary to convince key tourist subjects that better market and economic results can be achieved through interconnectedness. Individual factors must be compromised with the aim of generating optimal benefits for the greater good. Collectiveness is surely one of the key challenges for all tourist subjects. The next challenge is without doubt arriving at a consensus in operations, strategies, communication and prices, etc. In the past, inter-company contracts have proved to be a very successful tool as each subject within the framework of a tourist destination works for the destination as a whole, from which their primary business activity also benefits. This can be achieved only by formulating a common marketing and operational strategy, where the subjects agree to focus on e-business as a primary marketing and communication tool.

The case Norwegian Trysilfjellet BA shows us that individual tourist subjects in destinations benefit from interconnecting with others and implementing a system for managing public relations (CRM) with the aid of information communication technology. By implementing a virtual organization, investments in information communication technology are becoming more and more acceptable for all and in the long term ensure the only competitive advantage for the tourist destination.

References

Baretje & Defert (1972): Aspects of economic in tourism, Paris, Berger Levraut:66.

Britten, A. (2002): E-tourism in England, A strategy for modernizing English tourism through e-business; ENGLISH TOURISM COUNCIL Tourism Technology:55.

European Commission (August 2004): Electronic business in Tourism: Key issues, case studies, conclusions; The European e-Business market watch; Sector report: No. 07-II.

European Commission (May 2004): Electronic business in Tourism: The quantitative picture: Diffusion of ICT and e-business in 2003/04; The European e-Business market watch; Sector report: No. 07-I.

Jenkins, C. L. & Tosun, C. (1996): Regional planning approaches to tourism development: The case of Turkey, Tourism management, No. 17:196.

Jonker, J. A. (2004): The strategic identification and integration of critical success factors to achieve international competitiveness for South Africa as tourist destination; Doctoral thesis, University of Pretoria:115.

Kovač, B. et al. (2002): Strategija slovenskega turizma 2002-2006, Ministrstvo za gospodarstvo, Vlada Republike Slovenije, Ljubljana.

Leiper, N. (1979): The framework of tourism, Annals of tourism research:86

Leskovar, R. (2000): Kakovost programske opreme v novi ekonomiji, Organizacija (Kranj), 33, (7): 491-496.

Meglič J. (2004): More knowledge for more tourism, Management, knowledge and EU: proceedings of the 23rd International Conference on

Organizational Science Development; Slovenija, Modern Organization:221.

Ovsenik, M., Ambrož, M. (2000): Ustvarjalno vodenje poslovnih procesov, Turistica, Visoka šola za turizem:143-151.

Sibila-Lebe, S. et al. (2003): Analiza obstoječih in razvijajočih se turističnih proizvodov v slovenija, Center za interdisciplinarne in multidisciplinarne raziskave in študije, Znanstveni inštitut za regionalni razvoj pri Univerzi v Mariboru:13.

Smith, S. L. J. (1988): The tourism product, Annals of tourism research, No. 21:33.

Yoon, Y. (2002): Development of a structural model for tourist destination competitiveness from a stake holder's perspective; Virginia Polytechnic institute and state university:43.

Marko Tkalčič, Matevž Pogačnik

Tourist Adapted Destination Selection

Abstract

In this paper we present a framework for a tourist adapted destination search. The typical scenario where such a framework can be implemented is a tourist wandering around and looking for some attractions, events or activities. They are equipped with a mobile device that can connect to the Internet and a geographical location service (such as GPS). The system helps tourists during their travels by informing them of attractions, events or activities they might be interested in and which are situated nearby. Imagine two tourists standing next to each other - the user who is interested in architecture receives a message that there is an interesting building just around the corner while a user who is interested in shopping receives information that the shop he is standing close to is currently offering huge discounts. The system relies on a database of tourist information that is maintained by the tourist content provider. This content provider trades in this information based on the business model. Many companies offering tourist services are interested in this new approach to promoting their activities. The database is a complex ontology based on a set of tourist information equipped with metadata about the location, type and rating of various tourist offers. The system also includes the tourist's personal profile, which is a data structure containing information about the tourist's likes and/or dislikes. The tourist's profile is built on the basis of his history of past selections and feedback on certain suggestions. Based on this data the system decides which information to deliver to the tourist's personal mobile device.

Key words: *destination, destination selection, destination search, data system, tourist profile*

1.   Introduction

It is a universal conception that a person strolling down the street with a paperback in his hands, a camera hanging around his neck and a backpack on his shoulders is a tourist. This stereotypical description embodies some of the most inseparable properties of a vast share of the tourist population:

- the tourist guidebook as a source of various items of information that may be useful to the tourist while wandering around and

- the camera as a tool that allows the tourist to make custom souvenirs to capture the personal experiences of the places they have visited

A tourist guidebook can be seen as a large database that stores information about a tourist location. Since guidebooks are sold in stores as an as-is object the publishers are trying to include information that can be useful to a large audience thus leaving the tourist to extract the information they need at any certain moment from the book . If we take a look at the table of contents of a normal tourist guidebook (Wilson et al., 2001:6) we can see that there is a number of chapters that some tourists are probably never going to use (e.g. Traveling with children, Senior Travelers, Gay & Lesbian Travelers). This redundant information has many consequences:

- the tourist loses time and patience while browsing and searching through the guidebook, which can cause them to miss some important information

- the guidebook is more voluminous than it could be, which has implications on printing, transport and storage expenses

Currently, searching for information is cumbersome. The amount of any kind of available information is growing at a tremendous rate, as is the number of its consumers. Today, searching for particular information on the Internet usually results in a vast number of hits, many of which are irrelevant to the user. It is unlikely that millions of users are so similar in their interests that one approach to an information search fits all needs. The main problem is that there is too much information available and that keywords, being the most common query mechanism, are rarely an appropriate means of locating the information in which a user is interested. Information retrieval can be more effective if individual users' interests and preferences are taken into account.

A similar situation prevails in the field of tourism, more specifically in the field of the tourist recommenders. Users are interested in different types of attractions: sights, events, museums, landmarks, etc., and their automatic selection for a particular user is not a trivial task. Furthermore, as most users travel in groups it is important to generate recommendations for groups of tourists.

The existing personalization techniques, such as content-based and collaborative approaches (see section 0), can be reused for these purposes. They should be used in combination with specialized user model structures, tailored for the field of tourism.

In order to have such a system for information retrieval we need to have the appropriate infrastructure. When designing the infrastructure we need to bear in mind that we want to reuse the existing infrastructure and applications for two main reasons:

- It is already there,
- The users are used to it.

This means that we need to use the existing fixed (telephone, cable, Internet) and wireless (GSM, GPS, GPRS, UMTS, WiFi) network infrastructure. We need to use standard middleware, hardware and software as well as standardized application level solutions (HTML, WML, XML).

Besides these technical requirements we need to have providers of content. Current tourist guidebook publishers and tourist service providers can put their content into the appropriate form (using ontologies, metadata standards and location information) that is searchable via the methods described in this paper.

All the players described above (tourists, technology providers, content providers, tourist service providers) can benefit from exploiting new business opportunities.

2. State of the Art

2.1 Existing Services

There are already several interactive web systems in the field of tourism. Michelin (www.via-michelin.com) offers a web service where the user can enter a starting and ending location (e.g. Ljubljana and Paris) and the service provides them with an optimized road route. The system allows the user to enter some constraints and displays also some additional data, like road toll costs.

Tomorrow Focus offers a web service that has lots of well-structured city tourist guides. The system is even available through some mobile operators.

Amadeus (www.amadeus.net) is a well-known web service that has a comprehensive database of flights. Based on the user's constraints (departure, arrival, dates) the service offers the user a set of available flights that match his criteria. Furthermore, the system has links to accommodation reservation services.

2.2 Basic Personalization Approaches

The field of personalized recommenders has a long history and is present in a number of domains such as web documents, e-mail, news, multimedia, TV programs, jokes, music, cooking recipes, tourism, etc. Although the approaches used are specifically tailored to the area of usage and employ different techniques, we can group them in a couple of main categories. These are content-based, collaborative, hybrid, stereotypical and demographic approaches. A good overview of their properties is made by (Burke, 2002:335).

In a content-based system (Linden et al, 1997:68, Billsus et al., 1999:333, Buczak et al., 2002:125, Ardissono Goy et al. 2003:688), the content items are defined by their associated features. For example, text recommendation systems such as Web page recommenders or news recommenders use the words of their text as features. In case of multimedia items, like videos, photos, or TV programs, metadata descriptions are used. These are structured descriptions containing descriptive fields such as genre, keywords, year of production, list of actors, etc. In the tourism sector these features can be a historical period, an artistic style, types of monuments or a geographical area, etc. A content-based recommender learns the profile of a user's interests, based on the features revealed by past items that the user has explicitly rated or implicitly used (for example, visited a web page about a particular site). The type of the user model derived by a content-based recommender depends on the selection method employed (word vectors, decision trees, weighted sum of attribute similarities, etc.). However, the basic process in the selection of appropriate content for the user is a comparison between the user model and the descriptive features of the collection of available items. All items in the collection are sequentially 'compared' with the user model, giving some kind of rating or similarity measure between the content item and the user model. The most similar items are considered the most appropriate and the top listed ones are recommended to the user. The advantage of the content-based approach is context awareness, meaning that a user's preferences can be explicitly expressed in terms of particular concepts, keywords, genres, etc. This is the advantage over the collaborative approach. Relative to collaborative filtering, content-based techniques also face the problem that they are limited by the features that are explicitly associated with the objects that they recommend. This puts these techniques at the mercy of the descriptive data available.

Collaborative filtering (Herlocker et al., 2000:100, Burke, 2002:335, Breese et al., 1998:42) is probably the most familiar and widely used approach. Collaborative recommender systems aggregate ratings or recommendations of content items, look for similarities (commonalities) between users on the basis of user ratings and generate new recommendations based on inter-user comparisons. A typical user model in a collaborative system consists of a vector of items and their ratings, continuously augmented as the user interacts with the system over time. The greatest strength of collaborative techniques is that they are completely independent of any machine-readable representation of the objects being recommended, and work well for complex objects such as music and movies. Unlike content-based systems, they are not interested in types, historical periods, genres, concepts, etc. that the user might be interested in, but instead look for similarity in choices/ratings of different users.

Stereotypical and demographic recommender systems aim to categorize the user, based on personal attributes, and make recommendations based on demographic classes or membership in a stereotypical group. The representation of demographic information in a user model can vary greatly. The benefit of the demographic/stereotypical approach is that it may not require a history of user ratings of the type needed by collaborative and content-based techniques, but it is also less accurate.

We must stress the importance of the feedback used for updating the user model. Namely, user interaction mechanisms provide basic observation mechanisms that conclusions about user preferences can be based on. Usually, these mechanisms are implemented in the framework of the graphical user interface (GUI). Related to user interaction mechanisms is the question of the type of user feedback. Users can either be asked to explicitly evaluate the suitability of a particular content item (a sight, museum, natural park, etc.), or the system has to make implicit conclusions about their suitability (analysis of search queries, amount of time spent on a certain page, etc.). A combination of both, the explicit and implicit approach, has also been used. One would normally expect better results from the explicit feedback approach since implicit feedback systems have to make decisions relying on incomplete and uncertain information. However, some authors report improved results using implicit feedback (O'Sullivan 04).

3. Existing Solutions in the Sector of Tourism

Although other fields of personalized content recommenders are much more developed, some approaches exist in the field of tourism. Three of them will be described in the following sections.

INTRIGUE (Ardissono et al. 2003:687) dynamically generates a multilingual tourist catalogue and recommends sightseeing destinations and itineraries by taking into account the preferences of heterogeneous tourist groups, such as families with children and the elderly. In order to increase the users' trust and to help them choose the preferred destinations, the recommendations are equipped with explanations addressing the potentially conflicting requirements of the group members. The groups are partitioned into a number of homogeneous subgroups, whose members have similar preferences. Each subgroup model is automatically generated by the system by combining information elicited from the user with stereotypical knowledge about the typical tourist classes considered in the domain. At the beginning of the interaction, the user is asked how many people are traveling together. The system also asks the user to distribute such people in subgroups, on the basis of their characteristics (age, background, interests, etc.). For each subgroup, a registration form is displayed to provide the system with information about the subgroup members. On the basis of this information provided about a subgroup, the subgroup model is initialized by matching its characteristics against the stereotypical information, and the best matching stereotypes are used to predict the subgroup preferences and its relevance, obtaining a user model of the group. Stereotypes used (Ardissono et al. 2003:687) have been defined by analyzing the tourist population from different perspectives: different stereotypes do not only describe tourists with different interests, but also different classes of capabilities; e.g., visually or physically impaired. The device-dependent user interfaces are generated by dynamically producing the content to be presented and by applying a standard, XML-based approach for its surface generation.

The Automated Travel Assistant, (Linden et al, 1997:67) is an implemented prototype of the model that interactively builds flight itineraries using real-time airline information. The system starts with minimal information about the user's preferences. Preferences are elicited and inferred incrementally by analyzing the critiques. The system's goal is to present "good" candidates to the user, but to do so it must learn as much as possible about their preferences in order to improve its choice of candidates in subsequent iterations. Through the progressing interaction, the travel agent learns more and more about the client's preferences. The travel agent learns for example that the client prefers to fly American Airlines, is somewhat price-sensitive, and has both hard and soft time constraints. Typically, client preferences can be complex and reflect complicated tradeoffs between cost of flight, airline, and departure time. An automated travel assistant system aspires to this sort of interaction, where the system provides information about available options and the user provides information about the quality of those options. The system's user model and thus the quality of the proposed options improve over time, ultimately resulting in an option that is acceptable to the user.

In (Marti et al., 1999:311) the authors are trying to mediate a museum visit with the aim of it being dynamically personalized and using the system as a base of learning in order to adapt itself to each new visitor. The authors' main hypothesis is that the spatial movements of the visitors inside the rooms are an important source of information used for developing an implicit user modeling approach. The authors classify visitors into four main categories: ant (long visit, sequential, complete, physically close to works of art); butterfly (half-term duration, selective, less sequential); fish (quick visit, superficial, away from the work of art) and grasshopper (short visit, with a few stops, non-sequential). The detection of the physical behavior of visitors within a museum, taking into account the time, movements and trajectory, can facilitate the acoustic mediation generated by the system. A prototype has been developed in which physical user modeling is proposed, managed by a neural network based on concrete examples from field observation, following the mentioned four categories.

As we can see, different recommendation approaches and objectives are employed in a relatively focused field of usage, such as the tourism sector. Nevertheless, in most cases we have to deal with evaluations of individual items, which can be very heterogeneous (sights, events, airplane flights, museum exhibits, etc.). In (Ardissono et al. 2003:701) there is an interesting discussion regarding the estimation of the user's evaluation of tourist items or attractions. Authors claim that the overall evaluation of each item has to take into account the presence of necessary properties, which a tourist attraction must offer in order to be recommended to the user. The techniques such as Multi-Attribute Utility Theory (MAUT) (Linden et al. 1997:71) are based on an additive evaluation of items, and calculate their overall utility for the user as the weighted sum of individual utilities carried by the attributes of the items (the weights represent the user's interests in the various properties). According to (Ardissono et al. 2003:701), this behavior does not fit the requirements of the tourist domain, where the presence of totally incompatible properties must downgrade the evaluation of items in order to avoid unsuitable recommendations. Therefore they suggest using such methods where a property receiving a very low score must have a drastic impact on the overall satisfaction score of the item by bringing it to the minimum level in this way.

4.1   Metadata Standards

Today, we face a number of different formats for the description of tourist information. Besides the Dublin core standard there are dozens of custom metadata structures as well as ontologies. The main problem is cross-platform interoperability. If the metadata and ontology structures are not standardized it is difficult to extract or merge information from different sources.

4.2   Location of Users

For an efficient site recommendation it is crucial to know the exact position of the user. The two most widely used methods are the GPS system and the GSM location service offered by some mobile providers.

4.3   Problem Statement

Our aim is to propose a framework, the requirements and constraints that encompass the following issues: the technology infrastructure, the way data are stored, selection algorithms and the presentation of data.

5.   Proposed Solution

5.1   Use Case

The interaction between the end user (the tourist) and the service provider (recommender service) must include some mandatory steps, besides the obvious search functionality.

The first step is the authentication of the user. This allows the service provider to uniquely identify the user. By knowing the identity[1] of the user, the service provider can apply the search filters that are most suitable for the selected user.

---

[1] Here we are not speaking about the personal identity of the user. Instead we refer to a virtual identity that stores only information relevant for this service.

The next step is to identify the way the user is connected to the service. There are two important parameters here:

- Client type and
- Connection type.

Both parameters have implications on the way the user and the service interact. Based on these values the service provider decides the kind and amount of data to be transmitted to the client. For example, a user with a PC and a broadband connection can receive full sized text, images and video while a user with a mobile phone can receive only limited subsets of this data.

The last step before the user can start using the service is pinpointing his location. This can be performed either:

- automatically,
- semi-automatically, or
- manually.

Automatic pinpointing is completely transparent for the user as his device uses some form of location technology to get its location (GPS, GSM, combination of both or any other technology) and transmits it to the service provider. The semi-automatic approach is comprised of two parts: the device recognizes the approximate location and the user provides more precise information about where they are. In the case of the third option, the user must enter his location manually.

After the user is successfully logged in (identified and located) he can start to use the service. The service is supposed to be a combination of push and pull methods. In the pull method the user first makes a query and the system responds with the appropriately filtered information. The push method means that the service provider automatically sends filtered information to the user without them sending a request.

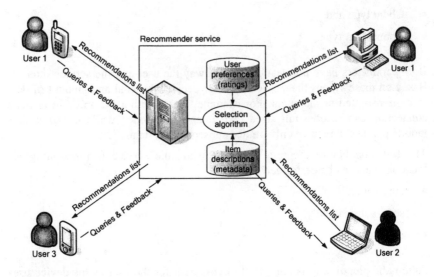

Figure 1: Use case diagram of the proposed framework. It requires at least one recommender service. Users can access the service via a number of different devices (PCs, mobile phones).

5.2   Personalized Recommender Service

The basic idea behind the personalized selection and recommendation of tourist sights, events, locations, lodgings, etc. (for the sake of simplicity we will call them tourist items) is relatively simple. The relevant recommender service has stored information about the preferences of many users and matches this against the descriptions of available tourist items. The first criterion for the selection is usually geographical, meaning that they are either close to the tourist's current location or the tourist has expressed interest in a particular region, city, etc. However, we should bear in mind that the selection of the target location could also be the result of the recommender service.

The items that match the user's preferences are selected and presented to his mobile device or PC, if the user is not (yet) traveling. In addition, some recommendations can be made based on positive ratings of other users similar to the target user, where the similarity was identified by analyzing past ratings (see section 0). The architecture of a personalized service in a tourist recommender system is presented in Figure 1.

The inclusion of the personalized tourist recommender service into the overall system imposes a number of requirements that should be discussed. They can be grouped into the architectural requirements and content-related requirements.

5.3 Architectural Requirements

In the proposed scenario the tourist uses his mobile device (mobile phone, PDA, etc.) while on the trip. The recommendations of sights, locations, restaurants, etc. are communicated to the tourist through the user application interface, which is also the entry point for the collection of user feedback. The information about user feedback regarding a particular sight, event or location is stored in a database. This database should, in our opinion, be kept in a centralized location and not distributed over the individual user's devices. This requirement is based on two reasons:

1. A single tourist user may have a number of different devices, through which he accesses the service. Most of the users who own a mobile device, especially PDA owners, most often have additional devices such as PCs, laptops, notebooks, etc. As the tourist item recommender service should be accessible from any type of device the feedback information is gathered from a number of devices. Therefore it makes sense to gather them in a centralized location.

2. Providing recommendations using the collaborative approach requires information on ratings of items from a number of users. As was explained in section 198 the collaborative recommenders calculate distances between different users and identify the 'closest ones' as the neighbors. Due to the well-known problems of collaborative recommenders sometimes referred to as the "ramp-up" problems (Burke 2002:340), their efficiency depends on the amount of user feedback provided by the users. More users also mean more information therefore centrally storing user feedback seems to be the most appropriate solution.

It may be argued that the information about user feedback and their preferences may be stored locally and later synchronized among the devices. Although this probably is a feasible solution, we believe that such an approach resembles a centralized solution as the information is gathered locally, only to be centralized at a later time.

5.4   Content-Related Requirements

If the tourist locations, sights, events etc. are to be selected for the user based on their appropriateness regarding user preferences, they must be appropriately described. As no metadata schema is considered a standard description in the field of tourism, a number of "custom-made" schemas are being used. This situation is no different from the situations in other fields (multimedia content, TV programs, music etc.) where numerous although occasionally compatible description standards are being used.

Therefore, we suggest the use of rules and employment of mapping techniques, providing some form of "translation" between the different description standards. In this way, the relations between different descriptive parameters (metadata fields such as the historical period, type of monument, cultural classification, ticket price, etc.) contained in the descriptions of different standards can be identified in a sensible way. What could be of interest is also the usage of ontologies, which can sometimes be seen as supersets of metadata standards. The relationships that ontologies contain may sometimes seem quite complex, but can provide a lot of useful information. They can on the one hand be used for presenting available tourist locations, sights, lodgings etc. as well as for more thorough analysis and identification of user preferences. A sample ontology for the tourism sector is presented in Figure 2.

5.5   The User Feedback Issue

The importance of user feedback is often neglected and over-simplified. Numerous algorithms work well by employing the database of rated items. However, the results presented by authors of recommender systems are often obtained in testing environments, where the users provide explicit feedback on the suitability of available items. In real-life scenarios the situation is not as simple as in the testing environments.

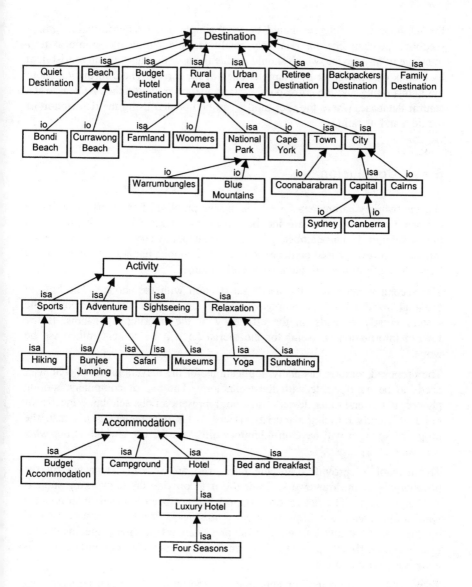

Figure 2: A sample ontology for the tourism sector

The average user is not willing to give much explicit feedback if any at all (Buczak et al., 2002:125) and expects the system to work well despite the lack of feedback. We believe that the users should provide explicit feedback, but that the methods for doing so should be very simple and straightforward. Simple feedback mechanisms are relatively easily feasible within the graphical user interface of any device, which should enable the user to rate a certain tourist item by a click of a button or by a simple touch of the screen. The information about the item's rating and the corresponding item identifier should then be sent to the central database, where the user model can be updated using the description of the item and its rating.

6. Conclusion

We see two key parameters for evaluating the proposed framework. The first is the share of tourists that are inclined to employ such a high-tech method of tourist support. If their number is not large enough, any large investments are not rational. However, based on the experience with mobile phones we foresee that a large share of tourists will favor such applications.

The second parameter is the user's satisfaction with the service. The biggest eventual drawback of the proposed solution lies outside the framework. The system greatly depends on the efficiency of the filtering algorithms. If the filtered information is useful for the tourist he will be content; otherwise he won't.

The proposed solution has many implications for the players involved. The user needs to be comfortable with the technology. The use of computers, mobile phones, PDAs and other devices must not represent an obstacle but a means for more efficiently utilizing a tourist service. If this requirement is not met, the proposed solution will have an inhibitory effect on the user, which is not what we wanted.

The technology providers are the ones who are severely impacted by the peculiarities of the framework. Their job is to provide the networking support and applications. The networking support has not been touched upon and the application layer is a delicate issue since most users decide on a product based on the ease of use and efficiency of the presentation layer, more specifically the user interface. How to present the information to users without inhibiting them is a big issue for the future.

Transcoding the existing contents into the appropriate form can be extremely time consuming. The application of ontologies and metadata is a lengthy but probably unavoidable step. Sooner or later content providers will need to create a well-structured database of their content in order to meet the new advances of technology.

Probably the biggest challenge for the providers is to find the appropriate business model that will allow them to fully exploit the possibilities currently offered by technology.

References

Ardissono, L. et al. (2003):. INTRIGUE: personalized recommendation of tourist attractions for desktop and handset devices. Applied Artificial Intelligence, Special Issue on Artificial Intelligence for Cultural Heritage and Digital Libraries. Vol. 17, (8-9):687-714. Taylor and Francis.

Billsus, D., Pazzani M. J. (1999): A hybrid user model for news story classification, Proceedings of the Seventh International Conference on User Modeling, University of California, Pattaya, USA:333.

Breese J. S., Heckerman D., Kadie C., Empirical analysis of predictive algorithms for collaborative filtering, Proceedings of the fourteenth conference on uncertainty in artificial intelligence, Madison, USA, Morgan Kaufmann Publisher, 1998:42.

Buczak, A., Zimmerman, J., Kurapati, K. (2002): Personalization: Improving ease-of-use, Trust and accuracy of a TV show recommender, In proceedings of the AH2002 Workshop on personalization in future TV, Malaga, Spain:125.

Burke, R. (2002): Hybrid Recommender Systems: Survey and Experiments User Modeling and User Adapted Interaction, Vol.12, 331-370.

Herlocker, J. L., Konstan J. A. Riedl J.: (2000): Explaining Collaborative Filtering Recommendations. Proc. ACM 2000 Conference on Computer Supported Cooperative Work, ECSCW'00, Philadelphia, PA:100.

Linden, G., Hanks, S. and Lesh, N. (1997): Interactive Assessment of User Preference Models: The Automated Travel Assistant. Proc. 6th Conference on User Modeling, UM'97, Chia Laguna, Italy:67-78.

Marti, P. et al. (1999): Adapting the Museum: a Non-intrusive User Modeling Approach. Proc. 7th International Conference on User Modeling, UM'99, Banff, Canada:311-313.

Wilson, N., Fallon S. (2001): Slovenia, Lonely Planet Publications, Victoria, Australia:6.

(Amadeus) www.amadeus.net.

(Michelin) www.via-michelin.com.

Emil Juvan, Rok Ovsenik, Goran Vukovič

Feasibility of Tourist Destination Management and its Development in Small Urban Areas – Case of the Mislinja Valley

Abstract

The demanding tourism market and its requirements are necessitating an ever more specific definition of tourism market offers. Planning such an offer demands an exact study of the feasibility of the development and the influences of tourism on the host environment (tourist destination). Destination management at this point meets the challenges of contemporary developmental trends and the pressure of local residents who want a stable, ecologically and socially focused and sustainable tourism economy. Throughout the prism of the third (service) level of the travel industry, we deal with new and contemporary trends of tourism development in the countryside and smaller urban areas, which arise from the fact that tourist destinations striving for success and long-term existence inevitably need strong internal connections. Only integrated operations and actions at both horizontal and vertical levels of the travel industry will enable small urban areas and small tourist destinations to follow the contemporary trends of tourism demand in the future. With the help of a tourism economy analysis in the Mislinja valley we tried to find possibilities for progress using new methods of integral tourism product development. Our findings could improve the existing situation of the tourism economy and help us identify opportunities for evolving new forms of tourism product development for a particular destination. The organization of global tourist destinations will be examined and some essential interventions into tourism development in the Mislinja valley will be attempted, which could consequently help to create an integral tourist destination - the Koroška region.

Key words: *tourism market, tourist destination, management*

1.   Introduction

Managing tourist destinations using the principles of integrated tourist destination management is a relatively new concept in Slovenia, especially from the standpoint of a theoretical, applicative, research and planning methodology. It is Ovsenik (2003:15) who first took this unique approach to the tourism management of Slovenian Alpine tourist destinations. The model that Ovsenik formed in his doctoral dissertation is especially interesting because Mislinja valley can be considered as an Alpine or sub-alpine destination. Nevertheless, this can be considered only as a starting point to be adjusted in terms of time, place, space and content according to the available tourism resources of the studied destination, which might be a very difficult task indeed. Changes affecting tourism will be the one and only parameter we will be witnessing (Ovsenik, 2003:55). Can we therefore hope for new challenges of creative research that will help us understand the development and influences of tourism on a local economy? These facts are pointing to a crisis in the tourism economy in small and undeveloped areas and regions. But these are not the only problems that tourism is facing today. To begin with, one of the greatest problems is the diverse terminology used in colloquial, written and technical communications in the tourism industry. Colloquially, travel agents are referred to as tourism agents, but considering their actual position between supply and demand we have to look at them as tourism brokers. This may seem an exaggeration, but misunderstanding the role of certain positions in the tourism economy may also result in the elimination of their roles. Ovsenik (2003:66) uses many different but also highly applicable terms for describing tourism economy organizations. By defining tourism at different levels he uses a variety of terms in an attempt to make them commonly used phrases. He refers to the basic processors of the travel industry as "travel organizers", which may sound strange but is the Slovenian equivalent to "tour operators" in English. On the other hand, there are contradictions in the tourism economy. Different tourism companies define "travel organizers" as "tour operators", which is a consequence of market relations. Tourism economy is international, so by using foreign phrases it is easier to do business in international tourism markets. The lack of educational institutions or their representatives on the one hand and the economy on the other may represent an important obstacle in developing successful models of tourism management in small urban areas. Planina (1997:45) defines a tourist destination as a juncture of tourist flows or sometimes even as "a purpose of tourism movement". He partially agrees with Leiper (1995:364) who defines a tourist destination as an element of the tourism system, as one of the elements of tourist flows. One trip or journey may have many different elements or destinations, but a tourist will start exploring those individual elements of tourist flows in his home town and they will eventually end their journey at the same spot.

As far as the Mislinja valley is concerned, it is still too soon to use the term "tourist destination", although it defines any area boasting a basic tourism infrastructure and thus stimulating tourist flows. We find many deviations and deficiencies in founding organizations that might assume responsibility for developing and promoting tourism offers. They may manage the information flow between subjects that are directly and indirectly involved in the system of tourism trade exchange. Those findings are certainly pointing towards an unsuitable tourism policy, which in many cases serves as a basic guide to certain important interventions and actions that need to be taken in order to develop a satisfactory tourism economy for a certain tourist destination (Baum, 1994:185). There are different levels of decision-making that need to be taken into consideration when we speak about tourism policies, whether European, national, regional or local.

2. Dimensions of Tourism in the Mislinja Valley

According to Khan et al (1993:49), tourism offers and their design are distributed among three different sectors in most countries:

- The government (as a developer and advisor),
- Non-profit organizations (as lobbyists and links between investors, local inhabitants and government), and
- Private capital (as investors).

These entities or links operate in different directions, which are, most of the time, contrary to the tourism policy. In the Mislinja valley there is no single organization (entity) that could act as a professional developer of a tourism economy or could be considered as a representative of one of the above mentioned sectors. There is not a single entity that could be either directly or indirectly engaged in the development of tourism or in the research of its influences on the host environment.

Table 1: Elements of tourism offer

Type of entity	Organization	No
Inns – pubs (places that offer a low diversity of prepared dishes and a small variety of à la carte dishes)	Private entity that can influence the development of integrated tour packages.	21
Museum – Art Gallery	Government/privately-owned infrastructure	9
Sports centers	Privately-owned sports infrastructure	25
Cultural Association	Association	35
Tourism Association	Association	5
Plants or elements of Natural or Cultural Heritage value	Managed by different bodies	34

Source: Juvan, E., 2003

The Tourist Information Center in Slovenj Gradec (capital of the Mislinja valley) symbolizes the alarming state of the quality of accommodation facilities in the Mislinja valley. The average rate is 2.2 of the maximum 5 stars for hotels, motels and other smaller lodging houses as well as 1.8 "apples" for accommodation facilities on tourism farms. Investment in a tourism infrastructure and superstructure is by far the most important issue for developing new forms of tourism economy for a destination (Law, 2002:8). What is even more important is that these activities be understood as providing significant impulses for the leisure and free time activities of both local residents and tourists, which may result in an improvement of the local economy. Law also asserts that such interventions promote destination as a business opportunity for different foreign investors. Gartner (1996:13) maintains that any environment or destination – provided there is enough interest on its part - can develop a successful tourism economy, regardless of the type of government. But at the same time he emphasizes the need for successful and open dialogue with the present local economy owners; otherwise the tourism economy might face unexpected and destructive consequences. This is even likely to happen when the government is the promoter of tourism development. Several researchers have emphasized the growing importance of the host community in recent years. It is widely recognized that planners and entrepreneurs in tourism must observe the views of the local community in order to develop a sustainable and long term tourism economy (Williams, Lawson, 2001:269). Of course, it is very important to predict the attitude of local companies towards working with the new external environment. In relation to this, we have to stress two main forms of corporate system behavior or their openness for external impulses. Ovsenik (2000:88) distinguishes "autopoetic" and "syspoetic" corporate systems, whose success very much depends on their internal or external business culture. "Syspoetic" systems are external and therefore likely to be more successful or even optimal for successful destination management. On the other hand, Roberts and Hall (2001:75) warn that it is very important to understand the individual role of every single element in the local tourism economy as far as tourist destinations are concerned. Especially today, when the demands in rural areas are not only more complex and varied, but are also on the rise. Opportunities are changing and very often shrinking (Butter et al., 1998:14). Government organizations and agencies act as supervisors of the development of a tourism economy, so their only task should be to enforce the national tourism development strategy. By insisting on that, we technically change the role of government and its bodies. The latter become accelerators or promoters of integration among the different tourism entities providing a certain destination.

Such a small number of tourism entities in the Mislinja valley may also be the result of natural selection in relation to the past economic capability of the environment. We must understand that tourism can never become the top income provider; nevertheless we must not neglect the fact that tourism can boost standards of living, and serve as a source of new jobs for local inhabitants and newcomers, as an additional source of monthly income for locals, as a solution for demographic problems, etc.

As the tourism offer of the Mislinja valley has been stagnating for several years, it does not generate sufficient economic results. This could be due to the lack of knowledge and expertise on the one hand and on the other hand because of insufficient investment funding. Considering the affirmation by Hall and Jenkins (1998:19), this might turn out to be an excellent opportunity for defining the active roles of tourism-oriented private, civil and government organizations in order to begin evolving a tourism economy.

There is no organized approach to modeling a tourist destination trademark or brand. A tourism trademark identifies itself as the most important element of a successful market presentation and market operation. As such, it can be characterized as a major competitive advantage of a tourist destination (Howie, 2003:71, Laws, 2002:8, Scott, 2002:187), (eco-tourism, uniqueness of environment, functional use of cultural heritage, futuristic tourism products, etc.). But only if destination branding depends on the genuine desire to establish good and known brands that will identify not only the needs of potential tourists but also their lifestyle. TIC (tourism information center) - the only government body responsible for information channels - is overloaded due to the abundance of work, and cannot even accomplish its basic task. The basic means of communication between demand and supply is still a classical brochure, (catalogue) which has informative value only, and does not represent a sales instrument. Public relations and marketing are both essential functions for a modern organization but in order to conduct successful marketing you need to develop a special organization structure that will meet the requirements of consumers throughout the organization (Cooper et al., 2005:704). Of course, the question of cost is not the only issue facing an efficient communicative instrument; the main problem is that of content. At most, only a few small companies are able to publish their own promotional and sales catalogues. The questions being raised are those of cost, knowledge and content as the companies do not have enough to offer.

Table 2: Promotional publications

Publication	Type	Publisher	Purpose
Slovenj Gradec and the Mislinja Valley	Monograph	Municipality of Slovenj Gradec	Records of the development of the Mislinja Valley
Slovenj Gradec and the Mislinja Valley	Misc-ellaneous	Municipality of Slovenj Gradec	/
Mislinja Valley guidebook	Tourism Guidebook	Municipality of Slovenj Gradec	Description of tourist attractions.
The Koroška region - Drava-Meža- the Mislinja Valley	Catalogue	Municipalities of Carinthia Region	Presentation of the three valleys
St. George – information catalogue	Catalogue	Municipality of Slovenj Gradec	Presentation of cultural heritage
By the homely fireplace – culinary imagery between Uršlja Mountain and Pohorje Hills.	Brochure	/	Presentation of culinary heritage of the Koroška region
Crafts from the Mislinja Valley (charcoal production, shingle splitting, gingerbread confectionery, candle making, shoe making, blacksmiths, etc.)	Brochure	/	Presentation of cultural heritage
Mountain tourism center Kope-Pohorje	Brochure	/	Promotional Brochure of the sport mountain center
Kompas Hotel Slovenj Gradec	Brochure	/	Promotional Brochure of the Hotel Slovenj Gradec
The Mislinja Valley and Slovenj Gradec invites	Brochure	/	Promotional Brochure of the valley
Slovenj Gradec - Tourism and cultural attractions	Brochure	/	Promotional Brochure

Source: Juvan, E., 2003

The basic question that arises at this point is the purpose of communication that does not achieve recognition or influence decisions of its targeted consumers and definitely does not contribute to sufficient tourism exchange (product or service for money). The web page of the town of Slovenj Gradec also covers the field of tourism, but the state of its home page and its value in terms of promoting tourism is poor. Due to the lack of knowledge about potential markets for tourism offers in the Mislinja valley and because the importance given to the media is low, there is practically no material published on the web, a major information warehouse where individuals or organizations can find a plethora of multimedia information and knowledge regardless of their location or ownership (Cooper et al., 2005:704). The offered services are basic and do not include reservations of tour packages. In consequence, promotion is severely compromised and the selling possibilities of tourism products via this most important and rapidly developing media are inexistent. Pütz (2003:115) states that more than 50% of all German trips will be purchased via the Internet by 2005, which is alarming for the Mislinja valley. The absence of Internet sales will push the valley to the very bottom of the competition rankings and it will lose a lot of tourism demand/income. Especially if we take into account that Germany accounts for the biggest share of the Slovenian tourism market as far as the flow of tourists is concerned. Ballantyne (2002:229) refers to London Business School forecasts and predicts that the number of all transactions between supply and demand will increase up to 60% by 2009, mostly in the field of tourism, travel and the hotel industry. There is a strong need to teach the new trends in the field of marketing communication and marketing in general. Reid (1998:188) maintains that destinations undergoing development should establish special government agencies responsible for the marketing of a destination. The effectiveness of such organizations was proven in Canada/Ontario, where the Ministry of Economic Development, Trade and Tourism supervised all marketing actions. There are different organizational schemes in place all over the world in the tourism industry. In France the government adopts a restrictive policy towards expenditure on outbound travels; the government of the Republic of China only recently opened its borders for foreign investors and is boosting outbound tourism. The government should mostly offer administrative and political support for developing the foundations of a tourism industry. If we identify a lack of knowledge in tourist destination marketing, destination management, and finally a lack of capital, then the development strategy should be based only on finding the expertise to improve the sales of the existing tourism offer. The income can subsequently be used to invest in new infrastructure and superstructure or to improve marketing communication channels. The requirements of income investments will decrease in some years and the local tourism economy will start generating financial profits. Devetak (2002:1037) warns that any development of a tourist destination should initially focus fully not only on developing infrastructure and tourism products, but primarily on researching tourism demand. This research must be sufficient, high quality and professional. Tourism needs and expectations are the basic element of this type of research. If there is a lack of knowledge in research and the methodology of marketing research, tourism companies should resort to educational institutions or professional marketing agencies.

The tourism offer in the Mislinja valley is dispersed; its individual parts (products or services) appear only as potential, but not as a real tourism product for they never allow a proper exchange. This spontaneous and unplanned tourism economy does not allow long-term planning or even the definition of any tourism strategies. Small family-owned tourism companies or those companies where tourism is not the main activity, in particular, do not sufficiently trust the government's strategies or help. This results in individual product development to enable the development or realization of any tourism strategies. With spontaneously developed tourism products travel agents (tour operators) cannot make tourism offers interesting enough, so they go somewhere else or they take over small family-owned companies and neglect the environment, economy and the demographic issues of a destination. As far as the level of tourism development, income and the average length of stay in the Mislinja valley are concerned, there are clearly many opportunities, but unfortunately nothing has been done yet. Vodovnik (2001:13) presents some alarming data about the number of tourists and other indicators of a successful tourism industry in the Mislinja valley in 2000. Hotel occupancy reached a yearly average of 9.8 %, with 22,806 overnight stays, which is low, and reveals an alarming situation.

The main issue at this point is the reason for such a situation. What is the state of the personal or business relations of the tourism entities in the Mislinja valley? Is it possible that in the near future we could establish a higher and more responsible level of cooperation among them in order to improve the present state of the tourism economy? Khan et al. (1993:49) maintain that very often the tourism companies themselves have to shoulder the lion's share of investments in tourism infrastructure. And if we agree with Gartner (1996:13) that the government should play the main role in uniting or joining the interests of individuals in the domain of supply and demand interest, then it is high time this actually happened.

The government bodies should help to overcome bureaucracy, evaluate the ecological consequences of tourism development, establish strategies for the sustainable development and preservation of the hosting environment, search for sufficient funds from the private sector and the government of Slovenia or the EU, and establish integrated forms of collaboration, etc. Developing tourist destinations must take into account every aspect of the magnitude of tourism demand, regulations and standards to ensure the destination's ability to withstand the pressure exerted on it by tourism demand (Ovsenik, 2003:66).

2.1 Human Resources

Vodovnik (2001:18) stated that tourism in the Mislinja valley employed 193 people in the year 2000. The average level of education is between 3rd and 4th levels. This data shows a need for an immediate and intensive improvement in the education and qualification of tourism staff. If tourism managers reach the 6th or a higher education level, then the available human resources require improvement. Tourism cannot be successfully developed without sufficient knowledge of tourist destination management. She stresses the importance of education at a higher applicative level and, above all, the use of new methods of teaching and tourism study for applied cases. That is why it is necessary to use all available forms and educational organizations at undergraduate and postgraduate levels, and all forms of permanent and vocational education. There is an unquestionable need for training and education at all levels of both the private and public sectors of the tourism industry (Cooper et. al., 2005:704). The available forms of education in tourism and institutions that offer this kind of education should be taken into account in deploying the strategy of tourism development in undeveloped areas (the Mislinja valley). Managers of tourism companies must recognize and develop elements that influence the effectiveness of tourism activities or the tourism economy in general. The tourism economy process must include experts who not only master the theoretical and technical approaches but also have the skills needed in tourism, but these people will also have to understand the point and philosophy of the service industry as such. By developing tourism offers they will depend on their own creativity. Of course, Cooper's pyramid can be used to understand the problem of the low average education level of employees in the tourism industry of Mislinja Valley. Cooper, Fletcher, Fyall, Gilbert and Wanhill (2005:704) claim that the tourism industry tends to yield a relatively flat occupational pyramid that can result in a small number of low and senior management posts. This often demonstrates a lack of career opportunities and consequently leads to a lack of staff motivation. Staff with low motivation levels will not render high quality service. Small rural areas are more likely to have small family-owned companies where jobs encompass many tasks, resulting in a shortage of opportunities for promotion.

The empirical findings of our research revealed the development of the following basic assumption:

H_0: In the Mislinja valley, the climate and level of culture among the tourism organizations and companies are not at a point where we could discuss a successful implementation of the modern model of tourist destination management into the studied environment in order to incorporate it into the global flows of the travel industry.

3   Methodology

By selecting the appropriate methodology of research we took the different elements of the studied environment (diversity of tourism offer, intensity of its connection to tourism, variety of levels of the economic development of the region, time needed to conclude the research, etc.) into account. We conducted the research personally because we thought that this approach might avoid a potential lack of cooperation on the part of companies and at the same time enable us to explain the importance of our research and its purpose for the local tourism economy.

We ought to point out that statistical research is not often conducted in this part of Slovenia. The pattern was random and the interviewees responded positively to the research. Following our inquiries, we collected statistical data using Microsoft-Excel and processed the same using SPSS software, version 10.0. We chose the method of descriptive statistics (variable, percentage and frequency) from the among the statistics techniques.

3.1   Instrument

The measuring instrument contains 17 closed-type questions split into two theme complexes (demographic and general tourism-related questions). Questions related to demographics information are descriptive. We collected data such as age of company, type of business, size of company, etc. Respondents were able to choose among a variety of different types of activities they might carry out, obviously in line with the actual registration of the company. In the second part of the questionnaire respondents give their opinion about some given statements or themes. The purpose of this part is to get an opinion about the researched tourism environment. Above all, we were interested in the relations and terms of local tourism companies towards the destination and tourism in general, the climate among different companies, level of competition, relations among the government bodies and the private sector, opinions about legislation in the field of tourism, etc. Respondents evaluate different statements on a Likert scale ranging from one (1) to five (5), where one (1) means "I don't agree with the statement" and five (5) means "I totally agree with the statement". Each question also has a short explanation.

3.2   Description of Pattern

We surveyed the managers of tourism companies and companies directly and indirectly related to tourism (as defined by the Standard Classification of Economic Activities in Slovenia). Among the respondents there were also some representatives of low management and directors of different civil and governmental associations dealing with tourism and tourism-related branches. We distributed 52 questionnaires; the analysis includes 49 of the returned and completed questionnaires. We included the entire Mislinja valley in our survey, including both municipalities (Mislinja and Slovenj Gradec). Since we were interested in the corporate opinion of the investigated issues, we decided not to ask for personal data of the respondents. We assumed that the respondents will be objective because they were told that we were only interested in the company's position towards tourism-related issues. In this respect we asked for demographic data of the companies. We examined two thirds of the tourism and tourism-related companies and organizations of the Mislinja valley. More than a third of them are classified as catering industry companies, followed by hotel companies. A small part of them are travel agencies, event companies and tourism farms. Non-profit organizations are represented on a rather small scale. Companies or organizations founded on the basis of civil initiative were tourism associations. More than a third of the companies and organizations involved in the research have been active for more than 15 years, almost a third for 6 to 10 years; others have been established more recently. Almost two thirds of the respondents revealed that they adopt an organized approach to submitting offers, and that they even conduct organized and integrated marketing.

4.   Results

Table 3 shows how respondents estimate the individual parameters of tourist destination management. 1 means *I don't agree with the statement*, 5 means *I totally agree with the statement*.

Respondents revealed positive opinions about the prospects of tourism income in the Mislinja valley. The average value was over three. If we try to take a deeper look into the results we can try to incorporate the results of similar research conducted by Florjančič and Bernik (2003:57). They discovered that managers of tourism companies who were involved in the research put emphasis on the following three elements of greater importance for tourism development: natural resources, human resources and marketing. So if managers of tourism companies in the Mislinja valley estimate that the Mislinja valley has an opportunity to develop income tourism, but do not have staff educated to a suitable level to perform the same, than this might be seen as an important sign for priority interventions into the organizational and managerial structure of the tourism economy in the Mislinja valley. Focus should be on education in tourism development. They also positively accept the idea of an integrated process of a tourism product and revealed the desire for integration with other companies. They expressed that they do take into consideration the wishes and desires of other tourism companies when they place a new product. Due to the fact that there are hardly any integrated tourism packages in the Mislinja valley, the average rate of 3.73 for the different models of cooperation with other companies more likely reveals that they do not focus only on integration and cooperation that would result in new tourism products. A standard deviation of 1.22 confirms that they are not exactly of the same opinion.

Respondents are pretty negative about the statement regarding a stimulating government regulatory environment. The average rate is only 2.09 out of 5. If we again try to compare the results of the research conducted by Florjančič and Bernik (2003:57) we realize that they had also discovered that 63.1% of their respondents thought that government regulation at national, regional and local level was not stimulating. Of the five options, the Ministry for the Economy and the National Chamber of Commerce ranked 4th and 5th (1st place was considered the best) in terms of their importance for stimulating tourism development. Considering these opinions we would expect increased activities in the field of integration at local level. We would expect new tourism development organizations to be established at local level that would increase the local tourism development initiative. On the other hand we can make assertions that these are the two most important fields that need to be regulated and organized. These opinions result from the weak decentralization process in the country. Representatives of urban and suburban areas do not have enough trust in the government and that is, of course, one of the main development obstacles. It is necessary to reduce the government's decision-making role by individual parts and elements of newly established organizations at local level adopting new development and creative approaches. The adequacy of human resources in the field of tourism is rated very low. It will be necessary to invest in human resources in order to raise the intellectual potential of new tourism strategies. An efficient educational strategy would definitely result in economizing the integration projects.

Table 3: Elements of tourist destination management

You are one of the tourism entities at your destination and you...	Variable	Standard deviation
think that income tourism represents a promising branch of industry in your environment?	3,89	1,05
think that tourism employees in your environment satisfy the needs of your local tourism economy?	2,39	0,95
take into account the desires and wishes of your local partners when creating new tourism products?	3,73	1,22
think that government regulation positively influences the development of suitable tourist destination management in your local area?	2,09	0,84

Source: Juvan, E., 2003

Table 4 presents the results and opinions in respect of the importance of integration and networking by individual, especially economically vulnerable small companies. Respondents evaluate the different elements of networking with parameters ranging from 1 (I don't agree) to 5 (I totally agree).

We encountered much respected opinions about networking. An average of 4.46 indicates that respondents show a positive attitude towards creating business relations with different business partners. The standard deviation shows a reasonable spread which is even more acceptable than with some other issues. The very same importance also applies to the opinion regarding collaboration between profit and non-profit organizations. Respondents evaluate this issue with an average grade of 4.67. This understanding of networking as an important element of success gives hope for new forms of business collaboration among small tourism companies on the one hand, but on the other hand it might also reveal the presence of other unexpected problems that have not been discovered yet. If this is not the fact here, we can expect that companies will soon begin integration processes in order to boost the diversity of tourism offers, empower marketing actions and finally improve the quality of services. However, the main reason for networking should not be profit gain and financial prosperity, but rather creating a diverse tourism offer that is ecology-sensitive and sustainability-oriented. If we imply that respondents do not trust the government and that they see legislation as a serious obstacle (Table 3) on the way to success, than the main initiative is expected to come from civil organizations and local companies. They will have to take the initiative to merge abilities and knowledge and to start creating effective networks to improve the local tourism economy. Of course, they will have to respect legislation, but at a highly applicative level. Slovenia is a new member of the European Union and therefore not so familiar with the different subsidies available from EU development programs. EU structural funds provide several billions of euros for investments in tourism infrastructure and organization processes. The government submits several public applications annually, but most of them are unsuccessful. The demanding procedures of application forms and processes are an obstacle, and since most small companies do not have access to any help in completing the application process, they ultimately fail to obtain any financial aid. The tendency toward collaborative processing in tourist destination that was revealed by our survey is grounds for hope that we might expect some improvements in the near future, but at the same time, it gives rise to new questions about when and what has not come of it so far. If representatives are aware of the importance of the tourism economy for the local environment, if they are enthusiastic about networking and are prepared to integrate with different profit and non-profit organizations, if they understand the importance of the diversity of tourism offers, but see the government's role and legislation as being hindering and perhaps even destructive, then they should put all the strength and knowledge into creating strong internal bonds within their local environment. If they are prepared to establish new forms of foundation then financing is not the huge problem it initially seemed to be. An average grade of 3.88 reveals a willingness to pay a special local tax for the development of tourism. But this issue is approached with caution, which can again be due to unpopular government policy. This mistrust and cautiousness on the part of the local tourism industry is particularly justified if the government continues to collect taxes and yet channels no or only very little money back into the local environment .

Table 4: Importance of business integration

Do you think that the networking of different tourism entities is vital for a successful operation of your and other organizations?						
Variable	2,00	3,00	4,00	5,00	no answer	total
Frequency	3,00	2,00	13,00	30,00	1,00	49,00
Percentage	6,12	4,08	26,53	61,22	2,04	100,00
Average variable value	**4,46**					
Standard deviation	0,85					

Source: Juvan, E., 2003

Table 5 shows the results of the analysis of opinions regarding brand imaging and destination branding. Respondents evaluate the statement with variables from 1 (I do not agree) to 5 (I agree).

Branding a destination image or generally designing one or more brand images of a tourist destination is one of the vital processes for establishing a competitive tourist destination. Creating attractive brand images definitely means an increased level of integrated operations among different tourism companies. The average variable shows that respondents evaluate this issue as vital! This is also one of the indicators showing a certain inclination to working together. Expectations of tourists when they decide to travel to a certain destination are based upon the image of the destination as a whole, not only of the hotel, the restaurant or similar elements. Consequently, this shows a conflict within the tourism system when there is dominance of one company (monopoly). Before opting for a certain destination, tourists will identify the quality, purpose and expectations of a trip or a vacation and will afterwards compare them with their personal needs and lifestyle. A destination brand image can be developed on a certain cultural or historical element of significant global value. If there is no such element, then we might face the problem of finding or developing an element that would fit so many different demands of the market. That is why it is very significant for any involved entity to identify itself with the destination brand; otherwise it will not be able to exist under that image. The standard deviation answer shows reasonable deflection. It seems that respondents could begin by developing a project in this direction. We presume that the tourism supply of the Mislinja valley does realize the importance of target marketing and destination brand imaging. Therefore, they should be thinking about establishing two main orientations in order to satisfy the needs for sufficient income and the needs and expectations of tourism demand. One such direction is tourism planning that would deal with:

- The tourism offer, tourism demand and destination management (expertise).

And the other one is tourism marketing that would deal with:

- Tourism product development, market segmentation and marketing planning.

Table 5: Possibility of creating an integrated brand

Do you think that there is a possibility of creating an integrated tourism brand for the Mislinja valley?						
Variable	2,00	3,00	4,00	5,00	No answer	total
Frequency	2,00	5,00	15,00	26,00	1,00	49,00
Percentage	4,08	10,20	30,61	53,06	2,04	100,00
Average variable value	**4,35**					
Standard deviation	0,84					

Source: Juvan, E., 2003

5. Conclusions

Tourist destination management in the Mislinja valley must implement steps that will ensure the progressive and successful adaptation of tourism suppliers, supply itself, the tourism infrastructure and the superstructure to tourism demand. Craftsmen that revive cultural history and historical lifestyles have managed to establish a link with the help of the center for development of small businesses and promotion of art craft, but their offer does not reflect that of the tourism market and tourism demand. On the basis of our research findings we can partially confirm and accept the basic presumption of the research. Results show that with the help of some intensive intervention into the organizational structure of the tourism economy we could manage to create conditions for implementing modern models of tourist destination management. Respondents attribute a big importance to networking, but do not explain why there are still no creative networks.

Tourist destination management here meets the challenges of mediating between suppliers, capital, government, local residents and tourism. It needs to develop in such a way to become a stimulating or initiating processor and booster of tourism exchange. But nevertheless we must be aware that the owner of tourist destination management (destination management organization) doesn't have an easy task. In order to create attractive tourist destinations they need to develop an appropriate tourism infrastructure and superstructure, they need to establish an adequate human resource policy and make plans about the use of natural resources and the land. Only then will they be able to create a tourism offer that will start generating profit from the whole process. Tourism suppliers need to be merged into interest groups that will critically evaluate their capacities and knowledge. They would then start to develop new tourism products in a range that fits the available resources of the destination in general. At that initial phase, developing tourism on a different scale would mean taking a different approach. The income generated by this approach can be used for investment, which is lacking in the present tourism economy. The region needs to establish a body that would act as initiator, counselor and developer for those interest groups and perform the vital tourism tasks of promotion, sales and service performance. Bear in mind that there are fields that are not covered at all (e.g. promotion, investments, administration support, etc.). Success is conditioned by the mutual linking of small tourism companies (Ovsenik, 2003:66).

Destination management organizations (DMO) in the Mislinja valley should address today's most neglected issues of the tourism economy (education, tourism development strategy, legislative support for small family-owned companies, marketing, research, integrations, etc.).

The development of a tourism economy in the Mislinja valley is left to spontaneous changes in the flow of tourists. In order to improve the present situation for a successful future it is necessary to engage in research. Research results will provide guidelines for the suitable creation of new tourism products and offers in general. These offers can easily be derived from demand (tourists) that will recognize their needs, lifestyle and personal beliefs either in new products or in a tourist destination itself. Market segmentation will help to apply the given resources of a destination to achieve sustainable tourism. Pohorje has a potential to fully develop a variety of thematic travel packages and other forms of leisure-oriented tourism products. The school center of Slovenj Gradec and the hospitality industry are developing simultaneously, but unfortunately separately from one another. If developments of both segments of the local economy could join forces with private capital, this could soon result in the foundation of a small congress center that would boost event and congress tourism. Essential improvements to accommodation facilities should be followed by the development of facilities that would raise the quality both of tourism vacations and of local leisure times. The awareness of local inhabitants plays a vital role and the development of tourism in a destination is very important, but unfortunately neglected too often. Our research displays that the interviewed managers of tourism and tourism-related businesses are aware of these facts and they agreed with the statement that local inhabitants play a vital role in tourism development. At the stage of development of a new tourism strategy and a new approach to tourism development, the local economy can benefit from interregional relations and experiences from their twinned cities in foreign countries and there is no need to learn from their own mistakes. Natural resources – longed for and pursued by zealous companies - are often endangered or a tourist destination supplements its natural attributes with cheap facilities, all for the sake of a diversified tourism offer. Of course, elements of tourism offers need to respect safety issues in every meaning of the term (health, feelings, conservation of environment, preservation of cultural heritage, etc.), but from the aspect of marketing they also need to be adjusted to targeted market segments and that often calls for an attractive appearance.

The establishment of destination information system (DIS) that will enable the transmission of information at the right time to the right place is crucial. Contemporary society puts pressure on the use of technology and computers, but in a specific environment such as the Mislinja valley, Pohorje, Kozjak, Plešivec, etc. the Internet won't reach these places anytime soon. In these areas the old-fashion panels, panel-boards and leaflets still look like the best solution. It is necessary to define the entry and exit points of a destination and when entering or leaving a destination a tourist must feel safe, self-dependent, that he is being subconsciously guided, and that his needs are important, etc. That is how a tourist feels safe and thus she/he will gladly explore the destination.

In closing, let us address the question of what to do in order to guarantee success. Is the lack of knowledge really the main reason for the current situation in tourism in the Mislinja valley or is it the low quality of accommodation facilities or perhaps the unfavorable policies of the government? Some answers were given in this paper, and others should be the object of research and developing projects still to come.

References

Baum, T. (1994): The development and implementation of national tourism policies. Tourism Management, 15:185-192.

Ballantyne, R. (2002): Investigating Consumer Choice in the E-Commerce Era: A Travel and Tourism Perspective. In Tourism Marketing: Quality and Service Management Perspectives. Laws, E. (ed). London: Continuum: 229-239.

Butler, R., et al. (ed.) (1998): Tourism and Recreation in Rural Areas. New York: John Wiley&Sons Inc.:14.

Christopher, M. et al. (2002): Relationship marketing: creating shareholder value. Oxford: Butterworth & Heinemann.

Cooper, C. et al. (2005): Tourism: Principles and Practice; 3rd Edition. London: Prentice Hall: 308 – 309, 704 – 709.

Devetak, G. (2002). Marketing v Evropskem okolju. In Management in Evropska Unija: zbornik konference z mednarodno udeležbo. Vukovič, G. (ed.).(2002). Kranj: Moderna Organizacija:1037-1080.

Florjančič, J., Bernik, M., (2003): Human Resource Management in tourism and the role of the Management Team. In 'Management v turizmu'. Jesenko, J. and Kiereta, I. (eds.). Kranj: Moderna organizacija:57- 76.

Gartner, W. C. (1996): Tourism development: Principles, Processes and Policies. New York: John Wiley & Sons Inc.:13

Hall, C. M., Jenkins, J. (1998): Rural tourism and recreation policy dimensions. In Tourism and Recreation in Rural Areas. R. Butler, et al. (eds.), Chichester, John Wiley:19-42.

Howie, F. (2003): Managing the Tourist destination. London: YHT Ltd.:71.

Juvan, E. (2003). Oblikovanje modela delovanja destinacijskega managementa v Mislinjski dolini: diplomska naloga. Portorož: Turistica-Visoka šola za turizem.

Khan, M., et al. (1993). VNR's Encyclopedia of Hospitality and Tourism. New York: Van Nostrand Reinhold:49.

Law, C. M. (2002): Urban tourism: The visitor economy and the growth of large cities. New York: Continuum:8.

Leiper, N. (2000): Are Destinations "The Heart of Tourism"?: The Advantages of an Alternative Description. In Current Issues in Tourism. Vol. 3, (4):364-368.

Mihalič, T. (2002): Ekonomika turizma 1: Program Turistica. Ljubljana: Ekonomska fakulteta.

Ovsenik, R. (2000). Model povezovanja potovalnih agencij v Sloveniji: magistrska naloga. Kranj: Fakulteta za organizacijske vede.

Ovsenik, R. (2003). Perspektive in protislovja razvoja turističnega področja, Model turističnega managementa na območju slovenskih Alp: doktorska naloga. Kranj: Fakulteta za organizacijske vede:15-66.

Ovsenik, M., Ambrož, M. (2000): Ustvarjalno vodenje poslovnih procesov. Portorož: Turistica-Visoka šola za turizem

Planina, J. (1997): Ekonomika turizma. Ljubljana: Ekonomska fakulteta:45.

Pütz, K. (2003): The necessity for Net Media in the tourism Industry: the increasing significance of the optimization of business travel with the Aid of Internet Technology. In "Management v turizmu". Jesenko, J. and Kiereta, I. (eds.). Kranj: Moderna organizacija:115 – 124.

Reid, D., G. (1998): Rural tourism development: Canadian provincial issues. In Butler, R., et al. (eds.). Tourism and Recreation in Rural Areas. New York: John Wiley&Sons Inc.:188.

Roberts, L. and Hall, D. (2001): Rural Tourism and Recreation: Principles to practice. CABI Publishing: New York:75

Scott, N. (2002): Branding the Gold Coast for Domestic and International Tourism Markets. In Tourism Marketing: Quality and Service Management Perspectives. Laws, E. (ed.). London: Continuum:178.

Vodovnik, P. (2001): Koroška... Tudi turistična pokrajina? In Koroška pokrajina. Močnik, V. (ed.). (2001). Slovenj Gradec: Cerdonis:13-18.

Williams, J., Lawson, R. (2001): Community Issues and Resident Opinions of Tourism. In Annals of tourism Research, Vol. 28. (2):269-290.

Zhang, Q. H., Chong, K., Carson, J. (2002), Tourism Policy Implementation in Mainland China: An Enterprise Perspective, International Journal of Contemporary Hospitality Management, Vol. 14, pp38-42.

Ministrstvo za malo gospodarstvo in turizem. (1998). Priročnik za izvajanje Zakona o pospeševanju turizma v praksi. Ljubljana: Republika Slovenija.

Ministrstvo za gospodarstvo. Predlog zakona o spremembah zakona o pospeševanju turizma. Avaliable on Oktober 25th 2003 on the Internet: http://www.mg-rs.si/datoteke/turizem/Predlog_ZRST.doc.

Bojan Vavtar

Supervision as a Function of Management in a Joint Stock Company in the EU in the Tourism Industry

Abstract

Corporate bodies have rights, entitled duties and obligations in legal relationships. In the theory of law, corporate bodies are natural and legal entities. Management is not a legal concept, which is why legal theory does not accept it from the standpoint of duties and obligations. This notion also does not use positive law. Therefore literally, the law does not define any rights or legal obligations for management. As a result, management in its general sense does not hold any legal responsibilities nor does it have any legal sanctions. However, the group of people whom we could call managers also without a doubt includes managers, members of corporate bodies and others whom we could define as managers. Their duty of supervision is defined within prevailing legal norms or originates from a professional responsibility. Both have an influence on the quality and responsible operations within the tourist industry. As a result of the nature of the tourist industry, the professional responsibility of management and with this its duty of supervision is especially significant. The common market must also be taken into consideration, which is becoming dominant in the tourist industry, and dictates the need for regulating European joint stock companies.

Key words: law, joint stock company, managing joint stock companies, management responsibility

1.   Introduction

The idea of a common market seems to be more realistic in the tourist industry than in the rest of the industries in the entire EU region. The idea behind the creation of a legal organizational formation within the economy, which would be subject to more than just national legal regulations, is as old as the common EU market itself (Merkt, 1992:652). In many cases a joint stock company is the legal form of a corporate body in the area of tourism. This status, like the formation of a company, would be especially appropriate in terms regulations that extend beyond the national legislature (Teichmann, 2002:409). The legal system of management responsibility directly impacts the quality of performing an activity in tourism (Vavtar, 2004:13). In an attempt to define legal responsibilities we come up against the problem of classifying responsibilities and, with this, defining the owner of these responsibilities. At first, it is necessary to separate overall responsibility into legal duties and those that are not of legal nature. We can define legal responsibility as the consequence of violating the prescribed treatment of a certain legal entity for which legal sanctions have been provided.

The characteristic of legal responsibility is the subject, who holds rights and obligations and a defined treatment or authorization of these rights and obligations. Usually, there are no problems with the first element in defining the responsibility of a subject when it includes a direct violation of the defined treatment. As a rule in positive law, the owners of rights and obligations are exactly defined. Problems arise in terms of how to define the content of their rights and especially their obligations, when they are not directly defined but originate from the responsibilities for a certain condition, achieved success or originate from the results of their business operations. We could also talk about indirect responsibility or a sample during treatment or the omission of treatment and its consequences, which have occurred as a result of this. A typical example of this type of treatment or omission is performing the supervisory function.

The responsibility of management in performing the supervisory function can be found in relation to its content and extent as well as within its legal qualifications. When performing supervision it is directly defined by legal norms and the decision. If supervision has been carried out in accordance to the regulations, it is dependent on each and every presumption of the actual treatment of the legal subject in relation to its obligations and its responsibility for its condition.

In most cases, although the obligations are clearly defined by legal acts as obligatory regulations, the responsibility - which may have created the certain condition - depends on the congruent alignment with the expected standards.

These standards of treatment are defined by the regulations of the profession and in the legal sense are expressed as legal standards.

The use of legal standards for measuring the treatment of civil responsibility is not usually problematic. There are more problems in applying legal standards to establishing culpable or subjective liability within the proceedings for establishing criminal liability.

In the following we are going to deal with the problem of management liability in performing the supervisory function in a joint stock company.

2. Defining Management in a Joint stock Company

2.1 The Management Board and the Supervisory Board of a Joint stock Company

Management of a joint stock company can be based on a one-track or two-track system of management. In both cases the management board is the managing body and leader in the joint stock company. A two-track system of management comprises both a management board and a supervisory board, which represents the owners of the joint stock company and supervises the company's managing board. In a one-track system of management, where only the board exists, the owners carry out the supervisory function by holding a general meeting or an extraordinary general meeting. Therefore, the law on corporations (The Official Gazette RS no. 32/93 and annexes – hereinafter ZGD) addresses both one-track and two-track systems for joint stock companies (Mežnar, 1995:714).

There is a definite need for corporate legislation governing the management of joint stock companies to become closer knit with national legislation (Korts, 2003:4).

2.2 The Managing Board in a Business Organization

Management is a general term for persons or bodies that manage the business operations in any type of business organization (Bohinc, 1996:338). According to ZGD these include persons who are legal and authorized under the law to manage their business operations.

The managing board has jurisdiction over all decisions, which relate to managing business operations except for those that have been withheld by the shareholders (supervisory board), by law or by a founding act (Bohinc, 1996:338).

As a function, the jurisdiction of the managing board must be understood as management as opposed to supervising, which is the jurisdiction of the partners or the general meeting of shareholders or the supervisory board.

The concept of company bodies is therefore a broader term than management. The managing board therefore relates solely to the management of business operations. Other bodies include (the managing board in a joint stock company, a manager in a limited liability company, all of the authorized partners in a limited liability company and a general partner in a limited partnership) the general meeting and the supervisory board.

2.3   Management in a Joint stock Company

The function of management in a joint stock company is much broader than its bodies. Uncontested functions of management are planning, organizing, managing and supervising (Meredith). There is no doubt that the tasks in a joint stock company are also performed by many employees within the company who are not members of the managing board or supervisory boards. Similarly, we cannot rank the employees who perform all or individual management functions amongst managers. Dividing up the leaders and heads is possible in terms of top and middle management etc.; we cannot, however, use this type of classification to define the owners of legal liability.

The passing of a decree on the statutes of European joint stock companies, and the directive relating to the participation of employees in management are simultaneously separated (5, 6). Employee representatives in the management organs of the company are therefore defined as part of management both in national legislation and in future European joint stock companies.

3.   Defining the Supervisory Function

Each and every treatment, which has been triggered by an individual decision, has a consequence. We make a decision and act on it precisely because we want to produce a certain consequence. If that were possible, everyone would lean towards the notion that their actions will only produce such desired consequences, therefore, we must in some way tie together the actions of the individual who is the decision-maker and the consequences that occur as a result of the actions (Meredith). Supervising is therefore a systematic function, comprising, above all, corrections to the system and not merely looking for mistakes made by individuals and allocating responsibility as a consequence. Compare (Referentenentwurf Gesetz zur Ausführung der Verordnung, 2001)

The concept of the management cycle includes analysis, planning, implementation and control, with the results from each cycle providing input for the next one (Council Regulation, 2001).

Supervision in both the private and public sectors is a fundamental element for successful management, which contributes to the effective management of an organization. Supervision is undoubtedly a function of management and at the same time there is a correlation between success in management and success in supervision.

In a broader sense, supervision is a sub-system whereby its function is to ensure a support system during operations. System interactions take place on the basis of information. A part of this information needed for managing the system is ensured by the supervisory sub-system. Efficient operations of the entire system depend on the quality of this information.

Supervision *sui generis* is an auxiliary activity. It consists of different procedures, which are characterized by their relative autonomy that is, above all, subject to objectivity. Objectivity is *sine qua non* effective supervision and at the same time the autonomy of supervision is limited by the supervisory function itself. A supervisory framework can only be defined at the abstract level, in advance and irrespective of the results of the supervision itself.

Efficient supervision is essential for successful operations of the organization, protection against negative influences within the organization itself and as an aid to realizing programs and goals. The supervisory function must therefore be included in the organization's system.

Last but not least, supervision is also in the definition of management.

3.1 The Supervisory Function in a Joint stock Company

In respect of supervising the management of a joint stock company from the standpoint of the administration board the law deems it the responsibility of the managing board to report to the supervisory board and the board's prescribed obligations about any losses, insolvency or financial embarrassment. (249, 256 and 257, article ZGD).

In relation to the general shareholders' meeting the board is obligated to prepare measures under the jurisdiction of the general meeting pursuant to demands, as well as contracts and other acts, whereby a consensus of the shareholders' general meeting is needed for validation and the performance of resolutions, which have been adopted by the general meeting under their jurisdiction (Meredith).

The management board must report to the supervisory board about the planning of business policies and other principal questions related to business operations, profitability, return on investment, the flow of operations, cash flow and the financial state of the company and operations, which can be very important for the profitability or solvency of the company. The supervisory board can also demand a report on other issues that are important for business operations. The report must be submitted to the supervisory board at least once a year and more than once if there are significant changes to the conditions. The company's balance sheets and income statements must be submitted to the supervisory board at least once a year during the discussion of the annual financial statements. A submission of reports on business operations, transactions, financial conditions, profitability and solvency must be made more often and thus quarterly. Reports about the company's profitability and solvency must be submitted each time an important business decision is made. Independent of these deadlines, the supervisory board can at any time demand a report on issues relating to the company's business operations and those that have a significant influence on it or can be expected to significantly impact the company's state of affairs (Meredith).

The managing board itself is obligated – if its preparation of the annual or interim financial reports reveals losses amounting to half of the share capital – to call a general meeting of shareholders and notify them.

In the case of severe liabilities or insolvency the board must propose the implementation of composition bankruptcy proceedings immediately or at least within three weeks of their occurrence, which the law defines as a reason for commencing bankruptcy proceedings. From the start of the company's financial insolvency or its severe liabilities the board may not execute any payments, except of course for payments that are compatible with the diligence of an honest and conscientious business person even after this time period. The law on the financial operations of a company is unambiguous when it comes to defining supervisory obligations and reporting.

All board members are subject to the obligations arising from the fundamental responsibilities for the payment ability of a company, not just the board member who acts as financial director. All transactions in respect of managing the company's business operations and realizing the duties of the financial function are not performed directly by the board members themselves. In order to successfully implement the financial policies an adequate information system is also needed, which gathers, archives and processes data and presents information relating to the implementation of financial decisions and the success of financial decisions for the financial decision-making process. Adequate accounting and analytical departments in the company gather and process this data. The board's duty is to ensure that the company's business operations are in accordance with the ZFPP and financial principles. This also includes the obligation to adequately organize the operations of these jobs in such a way that the data meets the requirements for simultaneous monitoring, planning and analysis (Plavšak, 2000:183).

In respect of the management board reports to the supervisory board, it is necessary to emphasize that the general shareholders' meeting takes jurisdiction in all cases where a company does not have a supervisory board.

According to the law itself, regardless of the company's form management shall:

- manage the company's business operations,

- make decisions about the business intentions of the company as an artificial formation, which the law grants legal capacity and therefore the ability to engage in legal relationships, conclude legal business operations and cause legal effects,

- represent the company and therefore express the business intentions formulated by management and other bodies within their jurisdiction in legal relationships and therefore legally bind the company and accept all rights on its behalf (Bohinc, 1996:338).

4. Defining the Term Liability

4.1 Legal and Non-legal Forms of Liability

Legal forms of liability include those types of liability for unlawful conduct of the owner, whereby the consequences are legal sanctions or sanctions, which are defined by the law. The sanctions defined by law depend on the level of social danger of the unlawful conduct, in this case illegal conduct, and are divided up into criminal liability (the sanction is punishment), liability for damages (the sanction is payment for the damages) and disciplinary (the sanction is a disciplinary measure). In accordance, with this we also differentiate between criminal liability for damages (also called material) and disciplinary liability (Bavcon et al., 2003).

We also rank liability for violating legal liabilities as a liability for a legally defined unlawful conduct with a lesser degree of social danger. Some authors also call this liability administrative or managing board liability. Besides the basic division of liabilities into legal and non-legal we also know the division of liability in terms of guilt and ownership of the liability. Within the framework of dividing liability in terms of guilt we differentiate between culpable liability (subjective) and non-culpable liability (strict). Within the framework of differentiating between the liabilities regarding who is responsible for the unlawful conduct, we separate this liability into individual and collective (Strohsack, 1996).

The above mentioned legal and non-legal forms of liability intertwine with one another in everyday life. If someone is convicted of a criminal offence they can also be perceived as being liable in a disciplinary proceeding for the same offence and as a result a disciplinary measure shall be given. If the same offence also resulted in damages, the individual will also be liable for damages and will have to pay for those damages. If someone is convicted of a criminal offence, which is in opposition to the morality and customs of a certain society or social group a non-legal liability will also arise.

4.2 Culpable (Subjective) and Non-Culpable (Strict) Liability

Culpable liability means that a certain individual is liable for unlawful conduct, if the conduct was caused with a certain form of guilt, i.e. if the conduct was intentional or negligent. Culpable liability is also called subjective liability because it is the foundation of the culpable liability as a subjective attitude of the individual towards the unlawful conduct. The preliminary condition for culpable liability is responsibility in law. Responsibility in law means only a person who is capable of understanding the consequences of his actions can be liable for unlawful conduct, which is why neither children nor the mentally impaired are responsible in law (Bavcon, Šelih, 2003).

Unlike subjective liability or culpable liability, non-culpable liability relates to the actual content and is liability for the consequences. The fact that the unlawful conduct of the individual has caused a forbidden consequence is sufficient. Non-culpable liability means that a certain individual is responsible for his unlawful conduct even if there was no guilt in the conduct or a guilty attitude towards the conduct, which caused the forbidden consequence (Strohsack, 1996).

Non-culpable liability is also called strict liability. The term strict liability, which is valid in the law of damages, above all, originates from the fact that substantial assumptions for the right to compensation are strict and not subjective. Damage and the causal relationship between the damage and the unlawful conduct committed by the person is enough for strict liability to exist, which is why it is also called causal liability. Strict liability or liability regardless of any guilt or liability according to the principle of causality was implemented into legislation mainly because of the liability for things or activities that represent an increase in danger (Strohsack, 1996).

4.3 Criminal Liability

The term criminal liability in layman's terms is used for denoting the liability of a perpetrator for a criminal offence. According to criminal law theory, the term criminal liability denotes one of the basic elements of a criminal offence. The term criminal liability is used in the narrower and the broader sense of the concept.

In the broader sense, criminal liability denotes the entirety of all the objective (man's free will, danger to society, unlawfulness and definiteness of the law) and subjective (sanity and guilt) elements of the criminal offence, which must be fulfilled for the perpetrator to be criminally liable and sentenced for the criminal offence. Criminal liability in the narrower sense of the concept denotes only a collection of subjective elements (sanity and guilt) of the criminal offence, which must be fulfilled in order to pass sentence on the perpetrator of the criminal offence.

In accordance with the above mentioned definitions of the term, criminal liability in the broader sense of the concept includes the existence of all the objective elements of a criminal offence and sanity along with guilt. The term criminal liability in the narrower sense includes only sanity and guilt.

The elements of criminal liability in the broader sense are:

- man's free will;
- dangers to society;
- unlawfulness;
- definiteness in the law;
- sanity and
- guilt.

In literature the term unlawfulness is denoted by the fact that the preliminary condition for criminal liability can only be the person's conduct, which opposes the legal order and basic norms of society. Definiteness in the law means that the preliminary conditions for criminal liability can only be the individual's conduct that is defined by the law in advance as a criminal offence. Sanity means certain psychological characteristics or denotes the fact that a perpetrator of a criminal offence can only be held criminally liable if he is able to recognize the world around him and is in control of himself (Bavcon, Šelih, 2003).

Sanity includes elements of willfulness and willingness.

The element of willfulness in sanity denotes the fact that the perpetrator was able to understand the meaning of his offence at the moment of committing it. The element of willingness in sanity denotes the fact that the perpetrator must have been in control of his conduct at the moment of committing the criminal offence. In literature, the term guilt is denoted as the subjective attitude of the perpetrator towards the committed offence in which the perpetrator was aware, or should have been aware of the fact that he had done something he should not have against criminal law (10,11).

The term guilt comprises two principle forms:

* intent and

* negligence.

In literature, the term intent is denoted as the greatest intensity of the perpetrator's psychological behavior towards the criminal offence. In respect of the level of intensity in this relationship we differentiate between direct intent and contingent intent. Direct intent is when the perpetrator was aware of committing the offence and had a desire to commit it. Contingent intent is when the perpetrator was aware of, and had consented to the fact that his conduct would produce a forbidden consequence.

The term negligence is deemed to be a milder form of guilt compared with intent and is divided into reckless conduct and negligence. Reckless conduct is when the perpetrator was aware of the fact that his conduct would produce a forbidden consequence and recklessly thought he could stop it or it would not even occur. Negligence is when the perpetrator was not aware of the fact that his conduct would produce a forbidden consequence, although his personal characteristics imply that he have been aware of this.

In the broader sense of the term, a special mutual relationship exists between both elements of criminal liability since the perpetrator's sanity or ability to recognize the world around him properly or to be in control of himself is a condition for being guilty. Whoever is insane cannot under any circumstances be guilty because they are not able to form such a subjective attitude towards the offence being alleged (Bavcon, Šelih, 2003).

Criminal liability as a liability for the committed offence as determined in a criminal proceeding.

4.3   Liability for Damages

Liability for damages is a liability for damages caused and the main intention is to ensure compensation for the unlawful conduct of the damages caused. Liability for damages does not cumulate with other forms of legal liability because it is autonomous and independent from the existence of other legal liabilities. Liability for damages is determined by the court in a legal proceeding, which depends on the disposition of the clients. Within the framework of liability for damages we differentiate between:

• contractual or business liability for damages and

• non-contractual or non-business liability for damages.

Contractual or business liability for damages occurs as a result of breaching a contract or violating an already existing legal relationship between a creditor and debtor with regard to a contract or any other existing legal relationship. Non-contractual or non-business liability for damages arises from encroaching upon someone's rights as defined by the law; the legal relationship between the creditor (the person who has damage claims) and the debtor (the person who caused the damages or the person who is liable) does not come into effect until the damage has been caused, and not before, which is the case for business liability. The subject of liability for damages (contractual or non-contractual) is the obligation to compensate or the duty of the client to compensate for the damages they are responsible for (Strohsack, 1996).

The main principle of the law of damages is that anyone can demand compensation for damages they have been caused within the following general conditions of liability for damages delicts:

• the damages are the result of unlawful conduct;

• the damages have actually occurred;

• a causal relationship exists between the damages and the unlawful conduct;

• liability exists for whoever caused the damage;

The listed elements must be given cumulatively. As a rule, if only one of the elements is missing no liability for damages exists. The term unlawful conduct as an element of the liability for damages denotes such conduct (commitment or failure), which in general is unlawful and unnecessary and would be separately prohibited or unlawful by legal standards. Damage as an element of the liability for damages denotes the decrease of some property (material damage) and prevents its increase (loss of profit) including causing bodily or psychological harm or fear (immaterial damage). The causal relationship as an element of the liability for damages denotes a relation between damaging unlawful conduct and the damage incurred. The damaging unlawful conduct and the damage incurred are in a cause-effect relationship. Liability for the caused damage as an element of the liability for damages denotes the fact that the person who caused the damage must be liable for the damage incurred. The liability of the person who caused the damage and the damage incurred is culpable (subjective, delict) or non-culpable (strict, causal and quasi-delict) or liability according to the principle of causality. Culpable liability is a rule in our legal system whereas the principle of causality is an exception (Mežnar 1995:714).

For culpable liability, the person causing the damage must be guilty. Guilt as an element of culpable liability is given if the person who caused the damages acted with intent or negligence. Intent by the person causing the damage is the highest level given if the person who caused the damage is aware of the consequences of his actions, allowed it (contingent intent) or wanted to (direct intent). Negligence (inattention) is when the person who caused the damage was aware or should have been aware of the fact that his actions would cause damage or recklessly thought that no damage would occur or that he could have prevented it.

Proven negligence is so called common negligence. The law of obligations also recognizes gross negligence, which denotes extreme inadvertency and exceeds common negligence. With regard to the intensity of the act it is close to contingent intent. Sanity is also a condition for culpable liability, since the ability of the person causing the damage to understand what is happening and take this into consideration influences his actions (Vavtar, 2004:13).

The essence of culpable or strict liability is that the person who caused the damage is already held responsible without guilt and purely because of certain circumstances on his part. In most cases because they are holders of harmful objects and perform harmful activities. In this case, the damage incurred is deemed to be the consequence of these harmful objects or harmful activities. The elements of strict liability are the harmful objects and harmful activities from which an increase in danger arises and thus strict liability of their owner. The terms harmful object and harmful activities are not legally defined but are defined by legal standards arising from court practice (Strohsack, 1996).

The main principle of the law of damages is that the person who caused the damage is responsible for the damage if he is culpably or strictly liable. In order to legally protect the person who has suffered the damage, a special institute exists for the liability of the other party in the law of damages. This category also includes the liability of legal persons for damages caused by employees to third parties, besides the liability for dealing with mentally ill or impaired patients, the liability of parents to deal with their children and the liability of others to deal with youngsters who are under their supervision (Strohsack, 1981).

When managing business operations, members of the management board must act with the diligence of a conscientious and honest business person and protect the company's business secrets. All board members are collectively liable for damages to the company that have occurred as a result of violating their obligations. If for example, there is controversy surrounding the honest and conscientious fulfillment of their obligations, they must prove otherwise.

Legislation states examples where members are obligated to compensate for damages (if, contrary to the law, investments are returned to shareholders; shareholders are paid interest and dividends; shares and the shares of other companies are registered, acquired, taken in pledge or withdrawn; shares are issued before the entire payment of the nominal or higher issued amount; the company's assets are divided up; payments are made after insolvency has occurred; the company generates severe liabilities; and capital is conditionally increased by issuing shares in contravention of the specified intention or before the entire payout of their value). In many of the cases listed above the board members are liable for damages if they break the law. They are not liable for damages if the activities are based on a resolution passed by the general shareholders' meeting. Liability for damages is not excluded in cases if the supervisory board approves the activities. Otherwise, the company can renounce the damage claim or reconcile it, but not until three years after the occurrence of the claim or if the general shareholders' meeting agrees and there is no written objection by the minority who together hold at least ten percent of the share capital.

The company's creditors can claim compensation for damages directly from the board members if the company cannot pay them back. These are cases where the board causes damages to the creditors as a result of their actions, and where the company itself, as primary debtor, is not able to compensate the damages caused to the creditors. A joint stock company is managed by the board for its own account (paragraph 1 of article 246 ZGD).

5. Conclusion

Liability for business decisions and obligations is regulated in different ways by legislation and includes aspects of economic, business, civil law – law on damages - and criminal law. These forms of management responsibility are also fundamental, both in other national legislations and in future European joint stock companies (Referentenentwurf Gesetz zur Ausführung der Verordnung, 2001).

It is possible to validly legally enforce a manager's liabilities as a responsibility in relation both to obligations, which are directly defined, and to responsibility for the conditions and business decisions. In both cases we also find liability arising from the omission of supervision. This can be directly defined by legal standards as the liability of top managers, liability where supervision is a consequence of not implementing responsibility for the state of affairs, or liability for business decisions. In the last case, particularly, there is generally no direct violation of prevailing laws but a breach of professional regulations, i.e. legal standards in dealing with a professional business person, a highly qualified expert, and as such, an indirect violation of legal norms. Therefore, a lack of supervision produces two consequences of management liability. Firstly, when the execution of the supervisory function is directly defined by law but not realized or implemented adequately, and secondly, when management performs its supervisory function in a manner that contradicts the expected standards of excellence and diligence. The need for careful operations in the tourism industry exceeds the norms of national regulations and the provisions regarding European joint stock companies. Supervision is therefore an important function of management, especially in companies in the tourism industry.

References

Bavcon, L., Šelih, K. (2003): Kazensko pravo.

Bavcon, L. et al. (2003): Pravo, Cankarjeva založba, Ljubljana.

Bohinc, R. (1996): Odgovornost poslovodstva gospodarske družbe Podjetje in delo, letnik 1996 (3):338.

Bučar, F. (1981): Upravljanje, Cankarjeva založba, Ljubljana.

Council Directive 2001/86/EC supplementing the Statute for European Company with regard to the involvement of employees; OJ L 294 (10.11. 2001): 22–32).

Council Regulation EC No 2157/2001 on the Statute for a European Company (SE); OJ L 294 (10. 11. 2001):1-21.

Jaeger (1994): Die Europäische Aktiengesellschaft – europäischen oder nationalen Rechts; Baden-Baden 1994:19.

Korts (2003): Die Europäische Aktiengesellschaft – Societas Europaea (SE) im Gesellschafts- und Steuerrecht; Heidelberg:4.

Merkt (1992): Europäische Aktiengesellschaft: Gesetzgebung als Selbstzweck; BB Š 1992:652.

Mežnar, D. (1995): Status uprave delniške družbe Svetovalec, letnik 1995, številka 21:714.

Meredith, M. MA (Cantab) MBA Dunelm MSc London ALI.

Plavšak, N. (2000): Obveznosti in odgovornosti članov uprave in nadzornega sveta Podjetje in delo, 2000 (2):183.

Referentenentwurf Gesetz zur Ausführung der Verordnung (EG) Nr. 2157/2001 des Rates vom 8. 10. 2001 über das Statut der Europäischen Gesellschaft (SE) –(SE-Ausführungsgesetz – SEAG).

Strohsack, B. (1981): Pravne odgovornosti, Center za samoupravno normativno dejavnost, Ljubljana.

Strohsack, B. (1996): Odškodninsko pravo in druge neposlovne obveznosti, ČZ Uradni list RS, Ljubljana.

Teichmann (2002): Die Einführung der Europäischen Aktiengesellschaft; ZGR:409.

Vavtar, B. (2004): The role of law in a strategic partnership in the area of tourism, Turistica, Portorož

Marko Ferjan

A Model for the Education of Tourism Workers

Abstract

The article presents an overview of curriculum theory. "Curriculum" is a relatively common word in initial education but one used less frequently in adult education, while "program" is more common. Some writers from Europe use both "curriculum design" and "program design" reasonably interchangeably, while American writers also use the term "instructional design". The concerns of adult education in the past have tended to be centered around the topics that have already been discussed, but without a great deal of explicit curriculum theory. Having analyzed theory and practice, the article explores their interrelationship. The second part of the article presents some ideas on how to put theory into practice.
Key words: Education, curriculum, staff, tourism

1.   Definition of the Problem

The paradigm of tourism activities has changed. In fact, completely new conceptual forms are emerging. (Ovsenik, Jerman, 2004:126). Many companies in Slovenia annually allocate as much as SIT 150,000 or more per employee for education and training. Of course, when so much is spent on education, the results of that education are very important. To be successful, the quality of planning, organizing, directing and executing this education must be comparable. A curriculum can be used as a good model of planning, organization, direction, execution and evaluation of the education.

The word "curriculum" has gained currency in Slovenia, especially since the 1996 changes to education legislation. The word "curriculum" is often

245

encountered in various contexts, but some also use the word "curricula". In my opinion, the interpretation of the meaning of "curriculum" is a textbook example of how Slovenes know how to import first-rate, but also completely unsuccessful, theories and practices from the wider world. We use the word "curriculum" in precisely this way: sometimes correctly, but sometimes completely missing the actual meaning. This is the first reason to shed some light on these problems.

2. Curriculum Theory Findings

The word "curriculum" is of Latin origin and literally means "course". The word "curricula" does not appear in the Latin-Slovene dictionary. In the 16th and 17th centuries the word "curriculum" meant "the ordering of instruction by year". One also frequently finds the notion of "curriculum" used with varying meanings pertaining to education. Many people understand "curriculum" as planning in education. In fact, planning is an activity, which allows management to systematically work (Ljubič, Ferjan, 2004:77). But the use of the word "curriculum" in educational theory originated in England.[1] Studying theory and practice reveals that this word (for many a difficult one to pronounce) has an unbelievable number of meanings, all of them relating to a system of education. George Posner (2004:5) states: "When various people discuss the meaning of 'curriculum', they imagine one of the following:

* Educational goals;

* Lesson content;

* Standards;

* Educational strategies."

A marked difference between these is that some have the final goals of the educational process in mind. It is about the contents, relations and experiences that a pupil gains at school. Others, however, have merely the course of the educational process in mind.

---

[1] In his book (Adult Education & Lifelong Learning, 2004, p. 273), Englishman Peter Jarvis, one of the authorities on lifelong learning theory, mentions a conversation with Malcolm Knowles, a very well-known American theoretician and author of many books in the area of education. Malcolm Knowles (1913–1997, called the "Father of American Andragogy"), is supposed to have declared: "What curriculum! A programme of course!"

Arieh Lewy,[2] one of the most important researchers in the area of education and human development who has written many books and articles in scientific journals, says that there are many definitions of "curriculum" that are mostly either evasive (and do not state what they ought to), long-winded or even completely mistaken.

A thorough study of English and American literature in this area, in particular, produces convincing evidence that this is indeed the case.

As Lewy (1991:13) reports, "traditionally, the interpretation of 'curriculum' has meant the entire body of knowledge or content that is taught at school." The formal document that doubly defines this is called a "syllabus" or a "teaching plan". The idea that curriculum is linked with long-term planning also comes from this sense of the word.[3] If one understands a curriculum as merely a "teaching plan", the basic question for schools in Western European civilization ever since the time of Plato has been: What should the plan include? (Kelly, 1989:30). John Dewey[4] was (to my knowledge) the first to introduce the expression "curriculum" to Anglo-American literature. In 1942 H. Harry Giles[5] and later in 1949 Ralph W. Tyler[6] both understood "curriculum" as **a process of teaching and learning**, which are nonetheless centered on the content. Their interpretation of "curriculum" is illustrated in Figure 1.

---

[2] He has published many books on this subject with UNESCO.

[3] More regular planning of learning materials.

[4] It is interesting to note that Pastunović (1999, 523) reports that it was 1902, whereas Jarvis (2004, 259) mentions 1916.

[5] Giles, H., H., McCutcheon, S. P. and Zechriel, A. N. (1942): Exploring the Curriculum, New York, Harper.

[6] Tyler, R., W.: (1949): Basic Principles of Curriculum and Instruction; Chicago, University of Chicago Press., cited in Kelly (1989, 15).

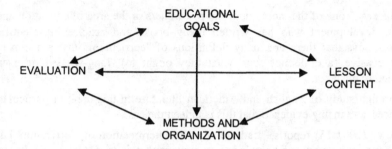

Figure 1: Tyler's curriculum model

For the sake of accuracy it is vital to very clearly emphasize that many writers in literature still treat the curriculum simply as the selection of lesson content, even though Tyler had presented a new model of the curriculum as long ago as 1949. He also conceived of the curriculum as teaching content, but he states that each curriculum must answer the following four basic questions:

1. What educational goals[7] do we wish to achieve?

2. What "educational experience" (qualifications) should the participants in education attain?

3. How should the "educational experience" be organized?

4. How do we ensure fulfillment of the purpose of the educational process?

In 1962, Hilda Taba[8] created a comprehensive definition of the concept of the curriculum. This concept includes:

- A declaration of the educational objectives;[9]

- Selection and organization of subject material;

- Methodological and organizational form of education; and

- Evaluation.

Lewy[10] however conceptualizes curriculum as a process of:

---

[7] He uses the term "goal".

[8] Original title: Taba, H. (1962): Curriculum development: Theory and Practice, Harcourt, New York. Cited in Jarvis (2004, 247) and Kelly (1989, 53).

[9] She uses the term "objectives".

- Defining the educational goals;

- Selecting the subject material;

- Planning the educational strategies;

- Preparing the educational materials;

- Engaging instructors;

- Evaluating the material; and

- Implementation.

Aside from proposed curriculum models there have also been many philosophical and other theoretical debates, which are of somewhat less interest to education organizers. However in theory, as already mentioned, "curriculum" is often defined as a long-term planning process in education. In this context some authors most often use the concept of curriculum to mean the process of planning teaching materials or simply the teaching plan, which includes preparing the **lesson schedule** as such. This concept of curriculum is not as complete as it could be, as stated above.

There are not as many inaccuracies written about the concept of **"curricula"**. The word "curricula" itself is also used less frequently. Indeed, it should be stated that even those that use the word do not know exactly what it means. From the context in which this word is used in literature one can conclude that the word "curricula" usually means "the course of schoolwork". It also had this meaning in the past; in the 18th century this is precisely what the word meant.

3. The Use of Curriculum Theory Findings in Tourism Education

Curriculum theory findings can easily be adapted not only to match the needs of educational institutions, but also those providing employee training. Lewy's curriculum model is already very useful for application at companies in its basic form and this is all the more true in a somewhat modified form[11].

---

[10] Cited in Lewy (1977).

[11] In his 1977 book, for example, Lewy merely speaks about the evaluation of materials. Of course it is necessary to conduct an evaluation along other lines as well, and thus it is not possible to merely speak of "evaluation of materials" or indeed "evaluation".

Analyzing curriculum from the perspective of process management functions shows that a curriculum already includes all of the process management functions: planning, organizing, leading and controlling:

- To a certain degree curriculum is already a plan in itself, because curriculum is used to define how and through what resources the education process will be conducted.

- In the curriculum process the function of the organization is also included when the organizational form of education is selected.

- In the implementation phase one moves from the level of planning and organization to the level of leading and performing when concrete tasks are assigned.

- Evaluation is a method of control in and of itself.

Indirectly, curriculum also includes motivating those participating in education. Research shows that both the content of education and the way in which it is executed are very influential on the satisfaction of adult participants in particular. Thus the motivation of participants is greatly influenced by the content of education and how it is carried out.

Figure 2 shows the group of activities in curriculum.

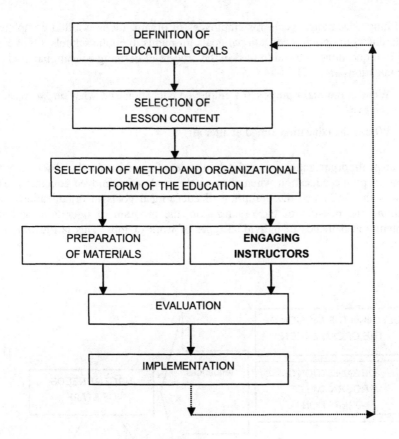

Figure 2: Collective activities in the curriculum

DEFINING THE GOALS OF EDUCATION in the curriculum process includes the following processes:

- Observing events in the social, academic and technological environment;
- Identifying educational goals that are still current and those that are already obsolete;
- Identifying the needs of participants in education;
- Giving final form to the goals of the educational structure.

Defining educational goals for employees in organizations is often completely different from defining educational goals for pupils in public schools. Of course, the first questions to be addressed in the process of defining educational goals in organizations are:

- What is the main purpose of employee education and what do we wish to achieve?

- Who is the education aimed at after all?

In for-profit organizations, completely different factors have to be observed from those in public education when shaping the goals of employee education. The factors that influence the definition of educational goals in organizations are: defining the mission of the organization, the program of operation, and the content and demands of individual types of work or jobs. This is illustrated in Figure 3.

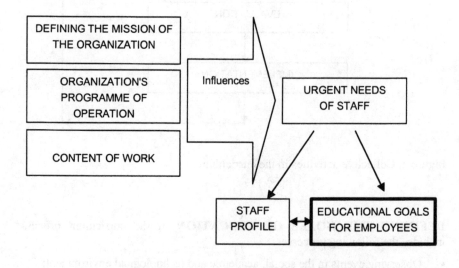

Figure 3: Factors that influence defining the educational goals for employees.

TEACHING MATERIAL FOR EDUCATION is selected to meet the educational goals. It is very important to take into account the following when selecting teaching material, particularly in relation to knowledge for informational purposes:

- Information about key facts and laws;
- Methodological knowledge of the specialized area;
- The dimension of specific values in the specialized area;
- The dimension of communication (specialized terminology, nonverbal forms of communication);
- The dimension of acquiring skills;
- Ensuring the ability to apply knowledge in practice;
- The possibility to build upon knowledge.

In practice, this phase is performed so that an inventory of the learning content to be imparted to the participants can be worked out in advance, followed by the development of a "learning content time schedule" that takes into account didactic principles (a systematic, step-by-step approach). This is especially advantageous because it must be possible to follow the educational process at every step. The time schedule for the learning material is the basis for both execution and control, and must also include dates for tests, for review and for possible grading.

THE METHODS AND ORGANIZATIONAL FORMS OF EDUCATION are generally chosen from among a number of alternatives. Therefore this phase of the curriculum is also often referred to as educational strategy planning. In this phase of the curriculum one must first consider which methods and organizational forms of education can actually ensure the achievement of the set goals.

Shaping educational strategies should be based on the following suppositions:

- The probability of achieving the educational goals;
- Economic justification;
- Adaptation to the needs of participants in education and users of staff;
- The availability of possible instructors and other resources for education;
- Participant activity;
- Participants' existing intellectual capital;
- Participant motivation.

The most important criterion for selecting a method and organizational form of education is the probability that the educational goal will be achieved.

THE PREPARATION OF MATERIALS includes:

- Preparation of teaching materials for attendees (printed and other materials);
- Preparation of instructors' resources (defining educational goals, preparation of content, additional sources of content, material for carrying out various exercises, etc.);
- Preparation of classroom equipment or the room where the education will be conducted.

In every case, the material must be prepared before the education actually starts. Traditionally, the instructor was seen as being the source of knowledge. This has been changing for some time. Access to information now means that technology makes it possible for both students and teachers to have more rapid access to teaching materials (the Internet). In preparing materials it is appropriate to keep in mind some additional rules or principles:

- The content of the materials must be oriented to the achievement of the goal established in advance;
- The content must be adapted to the background knowledge, interests and needs of the participants;
- Individual elements (e.g., textbooks, chapters, published databases) must contain only as many elements as do not overburden communication;
- The content of the material must not contain unnecessary elements (e.g., unnecessary descriptions, pictures, diagrams, backgrounds in poor taste, etc.);
- It is necessary to pay attention to certain phenomena connected with color: (e.g., contrasts, tediousness of colors, reaction times, popularity of colors, etc.).
- Etc.

Montilva, Sandia and Barrios (2002:205) even divide the preparation of materials itself into two levels:[12]

- The management level;

---

[12] Their paper applies to the preparation of material for publication on the Internet.

- The operative level.

The tasks of management in preparing materials are:

- Planning and control;
- Selecting tools to aid in the preparation of materials;
- Organizing the work of the team: the preparation of educational materials normally involves work that demands multidisciplinary knowledge: on the one hand, knowledge of the particular profession, and on the other knowledge of information technology and familiarity with programming tools; thus, teamwork is often necessary;
- Ensuring quality, which encompasses defining quality, quality control and all other necessary activities.

In contrast, at the operative level, it is especially important to pay attention to the following:

Table 1: The process of preparing educational materials. (Sandia and Barrios, 2002:208).

PHASE	STEPS
(1) Analysis of the educational content	Analysis of the educational field.
	Analysis of the educational goals.
	Analysis of the educational methods and organizational form.
(2) Analysis of the needs for preparation and distribution of materials.	Functional needs analysis.
	Defining requirements for mutual interaction.
	Defining requirements for further development.
	Defining requirements for quality.
(3) Definition of the content and its structure.	Planning structure.
	Planning content.
	Building the test version.
(4) Direct preparation of the materials.	Arranging the individual components into the whole.
	Final form of the prototype.
	Evaluation prototype.
(5) Distribution.	"Tidying up" or "supplementing" the prototype.
	Distribution.
	Final verification and delivery for use.

ENGAGING INSTRUCTORS to carry out the education comprises the following phases:

Defining the necessary qualities of the teacher means that the qualities that make up the personal and professional profile of the teacher must be defined. A teacher must have the right personal qualities in order to work successfully with people, so this issue must be addressed first when defining the profile of a teacher. Of course, the teacher must also have an appropriate professional profile, so it is vital to ask the following: What professional qualifications are absolutely necessary for a teacher and which are not? What professional qualifications of a teacher are we prepared or able to pay for? Defining resources means asking where teachers are sourced. Teachers may be obtained within an organization or from the outside environment. As for invited candidates, a final agreement must be reached with those that carry out education on how it is executed.

IMPLEMENTATION refers to the execution of what is planned or its realization. A proper course must be ensured with regard to implementation:

- Administrative and technical matters, which also include logistics and contact with learners and teachers;
- The teaching process itself.

All administrative and technical work must be planned, organized, led and monitored from the preparations to the conclusion. Here the following must be anticipated:

- The content of all activities to be carried out;
- Those that will carry out individual activities;
- The time period for carrying out the activities;
- Costs.

For the sake of transparency it is advantageous to plan these activities in the form of tables. Contingency plans to meet a "worst-case scenario" during the execution phase also have to be made.

The teaching process comprises the following processes: the planning process, the articulation process, the verification process and the conclusion. At this point only the execution or articulation is of interest. The **articulation process** (Cenčič, 1986:156) is the analysis and shaping of individual tasks within one teaching unit, and in connection with this the shaping of the teaching process as a meaningful whole.

Two things come to mind when speaking of articulation:

- Segmentation as part of the entire content of education that is defined in the schedule of the teaching material;
- The segmentation of one teaching unit.

The degrees of segmentation of a teaching unit are:

- The introduction, which consists of: an introductory greeting to the participants, announcing the teaching material, establishing background knowledge and motivating the participants;
- The core, where new teaching material is presented;
- The conclusion, which consists of: a summary, an evaluation, announcement of the next teaching unit and a farewell.

Homework is an important component of every teaching unit (for both young people and adults). The teacher assigns homework to the participants as well as checking the results of the homework. The purpose of the homework is to assign work to the participants in the educational process for them to become consciously active. They will only acquire long-term knowledge through conscious activity.

The **EVALUATION** is an element of every curriculum. It involves assessment. Assessment is conducted before the educational process as well as after it. Evaluation is conducted after every item in the curriculum. At the end, of course, the success of participants is also assessed. However, because there are many models of evaluation and methods of evaluation, it would not be fitting for such an important element of the curriculum to be presented cursorily. Evaluation will therefore be presented in a later article.

4. An Example of the Use of Curriculum Theory in Practice

In recent years travel has become an important part of leisure time. People are traveling more and more, and travel agencies are also offering a very wide variety of trips and destinations. These programs of course require people that are prepared to lead such trips. In Slovenia, as in other countries, this has led to a tour guide licensing system. In line with the regulations in force, tour guides and those accompanying tours must receive a license from the Chamber of

Commerce and Industry of Slovenia (CCIS). People taking the exam must have at least a secondary technical education (a four-year secondary school program, designated "Level V" in Slovenia) and knowledge of at least one foreign language at secondary school level. Candidates that have completed a three or four-year degree course are granted no special privileges on the exam. The same applies to candidates that have completed a secondary school program, or three or four-year degree course in tourism or a school specializing in tourism.

The remainder presents an example of an educational curriculum for preparing for the tour guide exam.

The **goal of the education** for which the model is presented is to prepare candidates to successfully take the theoretical and applied parts of the exam and, of course, to prepare the candidate to actually lead tours.

OVERVIEW OF AN EXAMPLE OF TEACHING MATERIAL SELECTION

MATERIAL	SHORT DESCRIPTION OF CONTENTS	LENGTH
Types and methods of leading tours	Introduction to the work of tour guides, terminology and differences between tour guides, essential characteristics and requirements of tour groups, variety of tourism programs.	4 hours
Hotel and airport procedures	Before arrival at the hotel, registration at the hotel, documents, staying at and checking out from the hotel, luggage, airline tickets, airline services, terminals, check-in, during the flight, lost luggage.	5 hours
Tour guide techniques on buses	Before departure, talking with a representative, receipt and preparation of materials, receiving guests, beginning the trip, technical aspects and contents of leading tours, concluding the trip, billing and report, other tour guide activities on the trip.	6 hours
Procedure for problem situations	Overview of the most common problem situations in tour guides' work and methods for avoiding them or resolving them, tour guides' responses during and after problem situations.	4 hours
Procedure at customs	Crossing the border with a bus, airplane, train, boat, tour guides' procedure at the border, forms for crossing the border, complications and resolving them while crossing borders, crossing borders abroad.	2 hours
Selected topics in guiding tours	City walking tours, tours using a local guide, touring buildings, tours using transportation other than buses and airplanes, specialized forms of guiding tours.	2 hours
The guide as representative and tour manager	Special characteristics of tour guide representatives' work at home and abroad, the work of a tour manager on lengthier, intercontinental trips.	2 hours

Psychology for tour guides	Overview of psychology issues for the national exam at the CCIS, psycho-physical qualities of tour guides.	3 hours
Public presentation	Proper use of the Slovenian language during a tour, guiding tours in foreign languages, using a microphone, speaking basics, characteristics of body language, mistakes while speaking, presentation exercises in the classroom.	3 hours
Etiquette and the guide's code of ethics	General etiquette, etiquette for tour guides, the importance of proper behavior while guiding tours, extracts from the tour guides' code (relations towards guests, the driver, partners, other tour guides, etc.).	2 hours
Tour guide literature	Presentation of the most important collections of guidebooks, other useful literature, printed matter, the use of CDs, the use of the Internet while guiding a tour (useful sites).	2 hours
Basics of operation in tourism	Documentation for tour guides, vouchers, contracts, the work of a representative, organizing and preparing a trip.	2 hours
Obligations	Overview of articles of the Obligations Act that are among the questions included on the national exam at the CCIS.	1 hour
Exercises and exam preparation	Concluding lecture with repetition of the basic topics of the seminar, instructions and preparation for the final simulation of guiding a tour on an applied, all-day trip in Slovenia.	3 hours
Presentation of tourism programs	Presentation of the most frequent one-day excursions in countries neighboring Slovenia from the perspective of a tour guide, slide presentation on guiding lengthier intercontinental trips.	4 hours
Tourism legislation	Overview of questions on the national exam at the CCIS with regard to various acts (the Promotion of Tourism Development Act, the Catering Act, the Frankfurt Table, etc.).	1 hour
Preparation of a seminar paper	Instructions and preparation of a seminar paper that the participants write in accordance with the instructions for a seminar paper for the national exam at the CCIS.	3 hours

Maps and their use	The use of maps in a tour guide's work, various types of maps, various scales, the most frequently used maps.	2 hours
Test and checking of answers	Written test with three questions and in-class review with comments on the answers.	2 hours
Topics in geography, ethnology, history and art history	Overview of questions for the national exam at the CCIS from the content areas mentioned, overview of literature for further study and preparation for the national exam.	8 hours
Applied section	One or two full-day trips in Slovenia with exercises in front of a microphone, simulation of guiding a tour, examination based on technical organization, content and speaking.	

OVERVIEW OF EDUCATIONAL STRATEGY SELECTION

The education of candidates is carried out in the form of seminars (lectures) and practical work. The education takes place from Monday to Friday for four school hours per day. One or two full-day trips around Slovenia are included, with exercises in front of the microphone, simulation of a guided tour, and examination based on technical organization, content and speaking.

OVERVIEW OF GUIDELINES FOR PREPARING MATERIALS

Participants in seminars should receive folders with the following work materials: the schedule and course outline, forms for crossing the border, and samples of various documents that guides need at work. They also receive the code of travel guide behavior, various travel and excursion programs, examination questions, literature and guidelines for writing a seminar paper for the national examination at CCIS, a sample seminar paper, official documents required for preparation for the examination, and possibly other things. The organizer prepares all these things and they are included in the price. Each lecturer prepares his own instructional aids himself.

ENGAGING INSTRUCTORS

The following instructors have been selected to carry out the education on the basis of their professional qualifications.

OVERVIEW OF EVALUATION EXAMPLE

The instructional goal is to prepare candidates to pass the examination at CCIS and also for subsequent high-quality work. The course content encompasses the thematic areas that are important for the examination and later in practice. Therefore, the course content is selected so that the participants will be able to meet the goals if they apply themselves well. The folder with materials for the participants has suitable contents. The course runs for a relatively high number of hours, which is optimal for comprehension of the course content. The organizational forms of the course are also chosen with learning habits in mind: they include both active and passive methods. The instructors have appropriate references. The place and time of the course can be adapted to the participants.

OVERVIEW OF GUIDELINES FOR IMPLEMENTATION

Activity	Leader	Period	Cost
Promotion of education	Person X	To ...	...
Collection of applications	Person Y	To ...	...
Renting space	Person X	To ...	...
Engaging instructors	Person Z	To ...	...
Administration	Person X	From ... to...	...
Preparation of materials for participants	Person Y	To ...	...
Trip organization	Person Z	To ...	...
Contact with CCIS	Person X	From.... to...	
...	...	...	...

EXAMPLE OF PREPARATION FOR CONDUCTING A TEACHING UNIT

The following is an example of instructor preparation for conducting the teaching unit "Etiquette and the travel guides' code".

Teaching unit phase	Content	Aids	Expected length
Introduction	Greetings, Taking attendance, Introduction of content, Assessment of previous knowledge	List of participants, Pencils, Copies of the code	20 minutes
Core	General etiquette, Guides' etiquette, Features of good behavior, Relationship with the driver, Relationship with guests, Relationship with business partners, Relationship with other guides	Overhead projector, Beamer, Laptop, Slides, Code of Behavior	80 minutes
Conclusion	Content summary, Question period, Work analysis	Same as above	20 minutes

5.   Conclusion

The use of known curriculum theories is especially suitable in adult education from the viewpoint of motivation. As we know, there are numerous unexamined motivational theories.

We certainly must not discount them. In my research in the past few years, however, I have determined that elements of the curriculum also especially influence adult participants in education, primarily:

- Course content;

- Educational materials (which have an especially bad influence on the atmosphere if we have not prepared educational materials or if the materials are poor);

- Instructor (primarily his level of expertise and ability to relate as a teacher to the participants in education);

- Execution or implementation views.

For these reasons I have determined that the use of known curriculum theories is also quite well-suited for educating employees within an organization as well as in tourism.

References

Cenčič, M. (1986): *Dinamika vzgojnega dela v šoli* (The Dynamics of Educational Work at School), DZS, Ljubljana:165.

Ferjan, M. (1999): *Organizacija izobraževanja, Moderna organizacija* (Learning Organization, Modern Organization), Kranj.

Giles, C. (1993): *Understanding Plans and Planning*, The Manchester Metropolitan University Centre for Education Management, Manchester (UK).

Gotz, K., Hafner, P. (2002): *Didactic Organization of Teaching and Learning Processes*, Peter Lang, Frankfurt.

Jarvis, P. (2004): *Ault Education and Lifelong Learning*, Routledge, London and New York.

Kelly, A. V. (1989): *The Curriculum: Theory and Practice*, Paul Chapman Publishing Ltd, London:30.

Knowles, M. (1998): *The Adult Learner*, Butterworth-Heinemann, Woburn, Massachusetts (USA).

Lewy, A. (1977): *Planning the School Curriculum*, UNESCO, Paris.

Lewy, A. (1991): *National and School-based Curriculum Development*, UNESCO, Paris:13.

Ljubič, T., Ferjan, M. (2004): *Planning in the tourist industry.* Published in: Jesenko, J., Kiereta, I. (eds.): *Management in tourism.* Frankfurt am Main [etc.]: P. Lang:77-100.

Montilva, J., A., Sandia, B., Barrios J. (2002): "Developing Instructional Web Sites-A Software Engineering Approach," *Education and Information Technologies*, vol. 7, (3):201–224, Kluwer Academic Publishers, Assinippi Park, Massachusetts (USA).

Ovsenik, R., Jerman, J. (2004): *Opportunities and Contradictions Involved in the Development of Tourist Destination: a Model of Tourism Management in the Area of the Slovene Alps,.* Published in: Jesenko, J., Kiereta, I. (eds.): *Management in tourism.* Frankfurt am Main [etc.]: P. Lang:125-172.

Pastunović, N. (1999): *Edukologija* [Educology], Znamen, Zagreb.

Posner, G. J. (2004): *Analyzing the Curriculum*, McGraw-Hill, New York:5.

Reynolds, W., M.: (2004): Curriculum, Peter Lang, New York.

Margareta Benčič, Rok Ovsenik, Iztok Purič

Controversies and Perspectives
of Destination Management

Abstract

For the establishment of destination management at Slovenian level we should act in at least two directions: upward from the local level – educating and raising the awareness of the local population – and upward from the regional and national level, when the division of force strategies in the operationalization of processes is the goal. This article provides special insight into the need for an altered role of the state in the sense of understanding and support of networking economic, profit and non profit interests at destination level.

Key words: *destination, management, education*

1. Theoretical Framework

Today we live in a time which is defined by uncertainty, competitiveness, greater complexity and less predictability. The explosion of knowledge, a modern telecommunications infrastructure and the presence of a culture of innovation have caused dramatic changes: the Asian economic crisis, the inception of Europe, mass takeovers, mergers and deregulation that have caused a new market environment, the Internet and the integration of supply chains to emerge. Therefore, we live in a time that is conditioned by constant change, whereby the position of individual sectors is significantly changing. The tourism industry significantly characterizes it as one of the most important industries in the service sector. As a result of rapid development and adapting to changes in the environment on one hand more open-ended questions are arising in the tourism industry and on the other hand there are more and more opportunities for the implementation of new strategies. Changes always depend on the environment, they occur in trends, which are usually faster than the changes themselves and the companies and organizations keeping up with the changes. In order to change the organization with respect to interaction with the environment, changes are necessary at two levels at least. Change within the culture and changes in the model of the organization (Burke, 2002:66).

The only constant that we can rely on, especially in the rapidly developing tourism industry, is the fact that there is continuous change and constructive destruction. It is sensible to deal with these changes as uninterrupted challenges and repeated deliberation of the "learning organization". Changes stipulate progressive globalization, the openness of markets and global competition. All the mentioned changes lean towards a digital revolution in communication technology (Champy, 1995:33). The consequence of this technology was a break in the bureaucratic organization and a break in the organization model according to the cooperative machine metaphor, which had experienced its triumph or zenith of internal organization for 25 years, observes Champy (1995:72).

Technological development inexorably demanded a restructuring in the nature of man's work within the national framework. In industry, because of robotics, human labor has almost become unnecessary, as the manufacturing process demands automated robots and knowledge workers. Human labor at lower levels (unskilled labor) is seen as a technological surplus. The new service industry in the tertiary sector of employment is seen as a new area and an opportunity for workers. Long-term change trends in the employment structure in the tourism industry have only been evident lately. A higher level of technological development is also demanded in the processes of the tertiary sector – service activities – more and more educated employees, that have to intellectualize the manufacturing processes

Qualitative features are increasingly replacing the quantitative features (Drucker, 1999:12), which is necessary when dealing directly with the service processes intended for the end-user. Processing also demands changes in the relations, which are consequently expressed in the changes of the structure. In the tertiary sector, where development and productivity lag behind, processing of the service is becoming a strategic problem.

Changes are necessary, which not only lean towards the transformation of the organization, but also towards changes in the culture. Changes create new dimensions of quality performance and a transition to obligation for a mission, the management of innovations and diversified management (Drucker, 1999:18). Attention is directed towards recognizing key trends, determining the influence of these trends, redefining the mission, the crumbling of the old hierarchy and building of the new one, and a flexible and free-flowing management structure, which releases the energy of the employees. The development of management does not have any characteristics that are innate to a country, but it internationalizes its weight towards universality (Florjančič, 1994:96). In its efforts and transformation towards universality and in relation to the country's traditions, in which it operates, management has some specifications and differences (Rehder, 1994:). As a result, the theory of management with the implementation and confirmation in individual systems becomes verified as a universally innovative system in all countries where changes are desired. (Florjančič, 1994:109).

Conditions for this include critical research of the work and processes, dispersion of responsibility, and communicating the mission, direction and values.

How can we anticipate the extent of change? If the dominant metaphor of the industrial age was a machine and a successful manager, then attention focuses on the system and the process. We could call the new model of change transition management; changes are external (politics, structure, and practical work), transformation is a psychological pre-orientation in people before change takes place.

Significant changes are rare, as they involve changes to the creative process with complete disorder, which is why companies must abandon continuity and direct themselves towards a mental level of decision-making processes. At any rate, the organization is dependent on the environment and trends in the environment cause changes. Adaptation and learning are demanded from the organization, if the organization wants to harmonize the organizational units in interaction with the environment. Even if the changes are old as the organization itself, they still demand a changed culture, in order to be able to change the mission and strategy. Throughout history we have noted different levels of change in organizations. From hierarchical organizations (Scientific management – Taylor) to changes that were dealt with by Hawthorn (who through organizational changes had already noticed that it is necessary to focus on individual and context variables in different ways) – group norms, expressed in the organizational culture, and system factors, expressed in the structure. We confront a system for an ideal strategic model for corporations by Blake Mouton with McKinsey who implements critical elements of professionalism alongside the builders of model organizations.

However, the objective of change is the system and not the individual. The model for understanding the life cycle of the organization structurally as well as a process (Capra) is researched auto-poetically, an independent core from the environment, which operates interactively through an open system (open structurally and closed organizationally) with the environment. The process is always a relationship between the imagined reality and the structure. That is why successful companies make decisions for internal changes through monitoring the environment and selecting significant changes that can be radical –

transformational (having discontinuity) or evolutional – transactional (continual). We always meet the builders of change at all the following levels: at the personal (participating in change), group (participating in teams) and organizational (realizing the changes) and at the level of effective management.

Different authors respond with different theories and approaches regarding the life cycle of changes and identify different elements as being vital: leadership, connections, being teleological etc. Maslow states that the life cycle of change directly depends on the motivation of employees, Hersberg adds hygiene to motivation along with enriching the work and Wroom and Lagler stress vocation and rewards. Hackman and Oldham state that feedback is a source of the results and aid in experiencing the significance of work, and Argyris stresses that the manager's integrity, approach and values are significant. Likert believes in participative management, and Lawrence and Lorch believe in all-out dependency, illustrated with relations: the organization and the environment, units in the organizations and members of the organization. Levison warns that in every organization the content of the changes is dependent on the processes, which add rational, dynamic and cognitive perspectives. He stresses that strategic changes demand mutuality in the content and the process and that they are not linear. He sees strategy as a method for implementing a mission.

The model of the conceptual survey of an organization (Porrass) signifies that it is a frame for the realization of a mission through the realization of a strategy. That is why the model of responsibility and competence is defined for the effective control of changes and achieving the mission through the organization's strategies on the basis of recognizable processes.

The model of responsibility and competence for the effective control of changes according to Nadler – Tushman shows the necessity that entrance into a system that is dependent on influences from the environment, sources and history is strategically organized and through process transformation shapes the type of organizations and groups in which the individual participates with great interest in the service. Tishy adds that this type of model is always dependent technically, politically and culturally.

McLelland adds that human needs, which are especially significant in the service sector, develop under strong influence from the environment.

Regarding the model of organizational results, Burke and Liwtin add that for the service factor it is especially significant in which kind of political and state supported environment the individual organization operates.

Therefore, the dilemma of organizational changes is put forth as a question; how to organize in a way that we will be able to improve our operations, find the right balance and within a dynamic environment do business in a relatively stable fashion with costs that can be foreseen and financial integrity protected.

As a result, a number of regions, especially in tourism, which we call destinations, have formed three typical organizational systems in their search for adaptation to change. These foundations can also be used in improving destination management.

The Burke Model

1) A hierarchical and bureaucratic organizational system of relations is offered for highly specific stable tasks. Tourism offers are often criticized as being bureaucratic – the result are precisely defined norms (legal enactments), which do not enable a flexible reaction to changes in the environment.

2) The market is seen as a coordinating organism with an exchange of specific, standardized capacities, which are subject to almost any fixed tasks e.g. costs amongst the carriers only represent temporary contractual agreements and minimal transaction costs. Only the information about the quality, quantity and market price is needed.

3) The strategic network can be an alternate cooperative mechanism, which is seen as a hierarchical scale or as the form of a coordination market. This can include dramatic changes of tasks. The dynamic tourism market is characteristic. The network consists of legally independent, mostly independently operating connected specialized organizations. This division of labor forms the structure of the network and small coordinating units at different levels are needed for its functioning.

4) The joining of people as an association; coordination is set up internally, and realized according to internal norms, values, quality and relations. Communication ensures that problems within the environment are solved quickly and efficiently, and in this way confronts the inflexibility of the environment.

2.   The Actualization of the Topic

The tourism industry is becoming a significant factor in the country's GDP. Tourism management has a significant role in the tourism economy, especially in developing tourist offers (Florjančič, J., Jesenko, J., Benčič, M., 1998). They even quote 70% of GDP, generated by tourism in the developed world. Throughout the world, tourism has created more than 44 million new jobs. It remains one of the most attractive industries for national economies and job creation. The strategy for further development in tourism should lean towards taking into consideration general economic laws that are valid for any region, of

Slovenia and beyond. On the other hand it should take comparative advantage into consideration, which would enable a superior position over the competition. (Florjančič, J., Jesenko, J., Benčič, M., 1998:96).

Therefore, tourism is becoming the dominating industrial method for human labor and management and is defined as: an economic activity that deals with the activity of satisfying tourist needs by offering services to them (SSKJ V, 1991). Whoever travels temporarily changes his place of residency for reasons of relaxation or entertainment and enables the service to be "hired labor", which is performed for payment. (SSKJ IV, 1991) (Eng. Service, Germ. Leistung). Service is defined as "someone doing something out of kindness, goodwill, compliance". (SSKJ V, 1991). Only having technical knowledge and skills is insufficient for performing tourism activities. Innovativity and creativity are becoming imperative for services in the modern service sector (Mundt, 1996:100). At any rate, the influence that a service has on a tourist can be direct or indirect. The service always depends on adequate organization (process and system) and the correct approaches taken by management. Therefore, a service in tourism is a multidimensional process, which includes many elements and creates a positive or negative experience for the guest.

Modern service-oriented companies incorporate three mutually connected groups of participants into the process, whereby the role is necessary for performing the service: guests, who have a need for a certain service, employees performing the service, who are responsible for the optimal performance of the service in the way the guests desire the service to be performed, and management, whose primary task is to coordinate the expectations of the guests and the experiences of the employees. All three roles are equally important and all three must be directed towards the objective.

When researching the above mentioned problem we were interested to see if this was the case in Slovenia. We propose two basic hypotheses to continue our research, which we are going to test.

U0:   Changes in receptive tourism cannot be restored by management for whom know-how is based on a mechanical approach.

U1:   Slovenian tourist organizations do not draw up strategic plans to incorporate themselves into global developmental currents in tourism.

3. Methodology

The population that was dealt with included managers from the tourism profession and the rest of the employees. We presented them with 13 thematic questions, which were formulated from different variables and influence each problem that was questioned.

We targeted the questionnaire from organization to organization. We were able to get 38 questionnaires from managers and employees from the tourism industry. They were proportionately dispersed across Slovenia and some of them were also from non-profit tourist organizations such as the Ministry of the Economy, STO, LTO, TZS and SZS.

For the technical part of the process we entered the data into Excel and processed it using the program SPSS version 10.0. We used descriptive statistical methods for the statistical techniques (averages, percentages, and frequency distribution), significance of change method (t-test for independent samples, ANOVA test and the χ^2-test) and the correlation method.

4. Description of Results

Chart 1: What region is your company from?

	number	%	cumulative%
Ljubljana and suburbs	7	18,42	18,42
Cerklje na Gor.	6	15,79	34,21
Bled-Bohinj	4	10,53	44,74
Škofja Loka	2	5,26	50,00
Koroška and the rest	12	31,58	81,58
No data	7	18,42	100,00
Total	38	100,00	

Source: Analysis of questionnaires

We also surveyed managers from tourist organizations in the region of the Slovenian Alps. We selected a random sample of 38 organizations located in the region of the Slovenian Alps.

Chart 2: Position of the company in the environment

Position	number	%	cumulative%
Leading	18	47,37	47,37
Following	9	23,68	71,05
Smaller	7	18,42	89,47
No data	4	10,53	100,00
Total	38	100,00	

Source: Analysis of questionnaires

From the results we established an interesting picture of the position of individual companies in their immediate and wider environment. Almost half of the companies who filled in our questionnaire defined themselves as being the leading company. A quarter of them defined themselves as a following and somewhat less than a fifth as a smaller company. Those who did not give an answer are probably associations and other non-profit entities in tourism such as The Tourism Association of Slovenia, The Ski Association of Slovenia and the Slovenian Tourist Organization etc.

Chart 3: Gender

Gender	number	%	cumulative%
Female	18	47,37	47,37
Male	17	44,74	92,11
No data	3	7,89	100,00
Total	38	100,00	

Source: Analysis of questionnaires

By random chance we carried out the research amongst almost completely identical samples of men and women, which is interesting because according to some statistical data the tourism industry employs a greater number of women.

A third of our respondents have a higher level of education and another quarter have university degrees or higher. Only a fifth of the respondents have only secondary school education. The relatively high percentage of those who have higher than secondary education is also interesting. Given that we looked for the respondents in the organization in a completely random way, the education level of our population is comparatively high.

4.1   Elements of Destination Management

We presented some basic elements of destination management development in the questionnaire, which had to be filled in order to discuss the modern forms of organization for individual subjects at a destination. The listed elements are also a basis for our hypotheses, which we had formulated:

Chart 4: Elements of destination management

Elements of Destination Management	Average Value	Standard Deviation
Do you feel that receptive tourism in your environment is a promising economic industry	4,03	0,90
When forming your complete range of services you also take into consideration the desires of the rest of the competition	3,73	1,07
Are you convinced that employee conditions are satisfactory for their optimal performance	3,21	0,93
Tourism products that you present on the market are prepared together with the rest of the tourist offers in your region	3,06	1,22
Do you feel that the personnel in tourism in your region have sufficient knowledge about their work	3,00	1,09
Strategic development of Slovenian destination management is adequate; aims are clearly defined	2,32	0,74
Do you feel that state regulations are formulated in a stimulating fashion and encourage tourist destinations to further develop	2,18	0,77

Source: Analysis of questionnaires

Chart 4 indicates how our respondents evaluated the individual elements using a scale from 1-5, where 1 meant I strongly disagree and 5 meant I completely agree. We also ranked them according to average value. The standard deviations of the results are also shown.

The respondents are united in their view that tourism is a promising activity for their region as the statement was supported by a value of 4.03 with a standard deviation of 0.90. The deviation was also greater for the statement: when forming your complete range of services you also take into consideration the desires of the rest of the competition. This indicates weak connectedness within the region and a low level of consciousness about mutual interdependence. They are also convinced that employees have satisfactory conditions for the optimal performance of their work – where the values are again 3.21 and the deviation is 0.93. The advantage of a response about the joint preparation of products for the market with the rest of the tourist offers had stopped at a value of 3.06, albeit with a standard deviation of 1.22 (the highest deviation amongst all the responses), which shows that the respondents are not united on this.

The poorest responses were given for the questions about the adequacy and target orientation of the company (only a value of 2.32 with a standard deviation of 0.74) and about the adequate stimulation from government regulations (2.18 with a deviation of 0.77). The results warn that there are two essential areas where matters have not been adequately settled – and even more – both prevent development.

Furthermore, we wanted to find out how are variables are linked with one another. We extracted the components using the method of primary components and rotated them using the Varimax rotation. We received three new factors, which explain the 72.59% total variance. At this point, the first factor explains 32.54% and the others 16.95%.

Chart 5 shows the factor structure.

In the first factor, elements, which are evident in the chart and explain the conditions of work, the level of education employees have and the prospects for receptive tourism were connected and that is why we named the new factor "The Prospects for Tourism and Tourism Employees".

For the second factor, elements relating to the products were connected, which is why we named the new factor "Products and Their Development".

The third factor unites the elements for the strategy of destination management development and the methods for state regulations. We have renamed the new factor "Development Strategies and State Regulations".

In our further research, we used the above mentioned factors as factor variables instead of the seven elements given.

Chart 5: Factor structure of elements

Factor Structure of Elements	1. factor	2. factor	3.factor
Are you convinced that employee conditions are satisfactory for their optimal performance	0,85	0,20	0,02
Do you feel that the personnel in tourism in your region have sufficient knowledge about their work	0,68	0,21	0,34
Do you feel the receptive tourism in your environment is a promising economic industry	0,68	-0,17	-0,48
Tourism products that you present on the market are prepared together with the rest of the tourist offers in your region	-0,02	0,89	0,01
When developing your complete range of services you also take into consideration the desires of the rest of the competition	0,42	0,77	-0,10
Strategic development of Slovenian destination management is adequate; aims are clearly defined	0,11	0,13	0,85
Do you feel that state regulations are formulated in a stimulating fashion and encourage tourist destinations to further develop	-0,04	-0,33	0,74

Source: Analysis of questionnaires

5.   Concluding Commentary

The discrepancies that originate from the research show that the situation is quite favorable with regard to understanding destination management and its significance including the quality of services. On the other hand, the participants evaluated the development strategy of Slovenian destination management as being less favorable, especially with regard to the clarity of the goals and their definition. The area of state regulations and their interference in the business operations of tourist companies were evaluated as being the least favorable. The respondents have established that state regulations are not set up in a stimulating fashion and do not motivate tourist destinations to develop further.

The state should clearly define the criteria, which must be sufficient for the leaders in destination management in order for tourist destinations in Slovenia to succeed in the desired way. Experience within the economy indicates that leading individuals must be aware of the specifications of modern destination management, especially in order to be able to understand its extensiveness. This task seems to be especially important because by using protective measures and subsidization after inclusion, Slovenia would be able to support the development of domestic destinations in order for them to rejuvenate themselves and ulitmately survive.

The state - including regulatory aims at local level - is portrayed as being a barrier. We see the local level as being the final completion of the destination design, which is being successfully managed by destination management. Destinations often unite many local communities and thus destination management must play the role of coordinator and middleman in order for it to operate in a balanced way. At local level, modern destination management should challenge the ideas, suggestions and impulses for innovation. This is why it shall also be necessary to reestablish a system of informing and educating the local population in order for the destination to operate successfully. Of course, in this case there are both prospects and discrepancies. There is the prospect of new jobs and a realization for everyone who will enter the tourist destination with more knowledge. The discrepancies show up in the expectations of local residents; that the destination will hire them regardless of their low level of education. Therefore, the set goals of every destination must first and foremost include informing the residents and employees about the vision and mission of the destination, including their strategic business plans. The service is processed with excellence in every phase of the process. The essential factor of the service is the individuals with their creativity, knowledge and personal characteristics, which crystallize into the most important element of first-class service, throughout the participation in the organizational process

Quality of service is one of the most important elements in destination management. Furthermore, effective destination management should recognize the prospects and discrepancies and react sensibly.

References

Burke W.W. (2002): "Organization Change: Theory and Practice", Foundations for organizational science, Thousand Oaks, California:106.

Champy, J. (1995): Reengineering the corporation: a manifesto for business revolution, London : Harper Collins:33-66.

Drucker, P. F. (1999): Management challenges for the 21st century, Oxford (etc.): Butterworth-Heinemann:12-18.

Florjančič (1994): Planiranje kadrov: Moderna organizacija, Kranj:96-109.

Florjančič, J., Jesenko, J., Benčič, M. (eds) (1998): Management v turizmu 2, Kranj : Moderna organizacija, 1998.

Mundt, J. (1996): Comparative politics: a theoretical framework, New York, Harper Collins College Publishers:100.

R. Rehder (1994): The education of the children of the Danish minority in Schleswig-Holstein, federal Republic of Germany: Strasbourg: Council of Europe.

SSKJ V (1991): Slovar slovenskega knjižnega jezika, Ljubljana, Državna založba Slovenije.

Jože Florjančič / Björn Paape (eds.)

Personnel and Management: Selected Topics

Frankfurt am Main, Berlin, Bern, Bruxelles, New York, Oxford, Wien, 2005.
492 pp., num. fig. and tab.
ISBN 3-631-53261-X / US-ISBN 0-8204-7390-1 · pb. € 74.50*

The book *Personnel and Management: Selected Topics* is a collection of various essays focusing on one the most topical problems in this field of research. The authors are well aware of the fact that this book presents only some individual topical chapters dealing with the issues of human resources and management. Nonetheless, they are convinced that this material can also have a significant impact on the deliberations about the development theory as well as its practical application in a given situation. The book has been published as the result of a long-term cooperation between the Faculty of Organizational Sciences of the University of Maribor, the Faculty of Economics of the University of Aachen and the European Centre of Integration Research (EZI) e.V., Aachen.

Contents: Management · Personnel

Frankfurt am Main · Berlin · Bern · Bruxelles · New York · Oxford · Wien
Distribution: Verlag Peter Lang AG
Moosstr. 1, CH-2542 Pieterlen
Telefax 00 41 (0) 32 / 376 17 27

*The €-price includes German tax rate
Prices are subject to change without notice
Homepage http://www.peterlang.de